T0140938

Ensemble Observability of Dynamical Systems

Von der Fakultät Konstruktions-, Produktions- und Fahrzeugtechnik
und dem Stuttgart Research Centre for Simulation Technology
der Universität Stuttgart zur Erlangung der Würde eines
Doktor-Ingenieurs (Dr.-Ing.) genehmigte Abhandlung

Vorgelegt von

Shen-Shen Zeng

aus Stuttgart

Hauptberichter: Prof. Dr.-Ing. Frank Allgöwer
Mitberichter: Prof. Dr. Lars Grüne
Prof. John Lygeros, Ph.D.

Tag der mündlichen Prüfung: 8. Juni 2016

Institut für Systemtheorie und Regelungstechnik
Universität Stuttgart
2016

Bibliografische Information der Deutschen Nationalbibliothek

Die Deutsche Nationalbibliothek verzeichnet diese Publikation in der Deutschen Nationalbibliografie; detaillierte bibliografische Daten sind im Internet über http://dnb.d-nb.de abrufbar.

D 93

ISBN 978-3-8325-4280-1

Logos Verlag Berlin GmbH
Comeniushof, Gubener Str. 47,
10243 Berlin
Tel.: +49 (0)30 42 85 10 90
Fax: +49 (0)30 42 85 10 92
INTERNET: http://www.logos-verlag.de

Acknowledgements

First and foremost, I would like to thank my advisor Prof. Frank Allgöwer for his encouragement and support, the invaluable scientific freedom, the highly stimulating research environment at the institute, and the opportunity to participate in numerous international conferences. I am also indebted to Prof. Roger W. Brockett for his kind hospitality during the three months I spent with him as a visiting researcher at Harvard University. I enjoyed and benefited from the many conversations that we had during my stay. I am also very grateful for the encouraging interactions that I had with the late Prof. Uwe Helmke and his group. Furthermore, I am very thankful to the members of my committee Prof. Lars Grüne, Prof. John Lygeros, and Prof. Arnold Kistner. Their interest in my work is greatly appreciated. Many thanks also go to my collaborators Prof. Christian Ebenbauer, Prof. Hideaki Ishii, and Jun.-Prof. Steffen Waldherr, for their encouragement, and valuable comments and suggestions. I want to thank all the members of the Institute for Systems Theory and Automatic Control for the very pleasant atmosphere, and the many interesting and fun interactions. Last but not least, I want to thank my family for their love and support.

Contents

Abstract

In this thesis, we introduce the concept of ensemble observability of dynamical systems and develop a theoretical framework in which this system property is characterized. An ensemble is a collection of nearly identical copies of one dynamical system, and it is said to be observable if the distribution of the nonidentical initial states can be reconstructed from observing the time evolution of a corresponding distribution of outputs. Similarly, a single dynamical system with output is said to be ensemble observable if a collection of copies of this system, with different initial states, is observable when considered as an ensemble. The consideration of ensemble observability is, in particular, motivated by recent efforts in the study of heterogeneous cell populations. Therein one aims to reconstruct a distribution of states within a population of cells, but is only given the time evolution of the distribution of certain measured quantities within the population. More generally, the motivation for introducing ensemble observability is rooted in the very concept of ensembles itself, in which a collection of individual systems may only be considered as a whole. A main result of this thesis illustrates a fundamental connection between the concept of ensemble observability and mathematical tomography problems. This insight is not only relevant for the considered ensemble observability problem, but provides a fundamentally new perspective on observability problems in general. Another main result concerns a natural connection to polynomial systems, which we encounter in the course of a systems theoretic treatment of the ensemble observability problem through the consideration of moments of the distributions. We also establish a duality of both approaches for linear systems.

Deutsche Kurzfassung

In der vorliegenden Arbeit führen wir das Konzept der Ensemble-Beobachtbarkeit von dynamischen Systemen ein und schaffen die theoretischen Grundlagen zur Untersuchung dieser Systemeigenschaft. Ein Ensemble bezeichnet eine Menge von nahezu identischen Kopien eines dynamischen Systems und wird beobachtbar genannt, falls die Verteilung von unterschiedlichen Anfangszuständen der Systeme im Ensemble aus der zeitlichen Entwicklung der Verteilung bestimmter Ausgangsgrößen der Systeme bestimmt werden kann. In analoger Weise wird ein einzelnes dynamisches System mit Ausgang ensemblebeobachtbar genannt, falls ein Ensemble aus Kopien dieses Systems beobachtbar ist. Die Einführung und Untersuchung des Konzeptes der Ensemble-Beobachtbarkeit ist insbesondere durch jüngste Betrachtungen von heterogenen Zellpopulationen motiviert. Ein zentrales Problem ist hierbei die Schätzung der Zustandsverteilung einer Zellpopulation, von der nur die zeitliche Entwicklung der Verteilung bestimmter Messgrößen gegeben ist. Allgemeiner ist die Grundidee für die Betrachtung von Ensemble-Beobachtbarkeit bereits im Konzept von Ensembles selbst verankert, in dem eine Menge aus individuellen Systemen nur in ihrer Gesamtheit aufgefasst wird. Ein zentrales Resultat dieser Arbeit ist die Herausarbeitung und Veranschaulichung einer grundlegenden Verbindung zwischen dem Konzept von Ensemble-Beobachtbarkeit und Fragestellungen in der mathematischen Tomographie. Diese Erkenntnis ist nicht nur für das betrachtete Ensemble-Beobachtbarkeitsproblem von Bedeutung, sondern für Beobachtbarkeitsprobleme im Allgemeinen. Ein zweiter zentraler Beitrag betrifft die Verbindung zu polynomiellen Systemen, welche im Rahmen einer systemtheoretischen Untersuchung der Ensemble-Beobachtbarkeit dynamischer Systeme durch die Betrachtung von Momenten der relevanten Verteilungen auftreten. Ferner wird gezeigt, dass die beiden Ansätze im Falle von linearen Systemen dual zueinander sind.

1 Introduction

Regarding a collection of individual systems as an *ensemble* is one of the most universal principles in the applied sciences. This principle has a long history, starting with its introduction in physics and leading up to its recent appearance in control theory, where it continues to act as a source of new fundamental concepts. Ensembles were first introduced in thermodynamics and statistical mechanics towards the end of the 19th century, with Boltzmann (1871), Einstein (1902), Gibbs (1902), and Maxwell (1879) being major contributors. Similar concepts were quickly developed and applied in other fields as well, such as in neurology in the form of so-called neural ensembles, in quantum control in the form of quantum ensembles, and more recently in biology within the study of cell populations, and in robotics within the study of robotic swarms, to name a few.

Even though the notion of ensembles does not always have the exact same description throughout these various disciplines, and, moreover, allows for different interpretations of the concept itself within one discipline, a rather general encompassing definition that fits well with the problem considered in this thesis is the description as an entity whose individual parts may only be considered together in a relation to the whole. Another perhaps slightly more specific and practical description for our purposes is that of a collection of nearly identical dynamical systems, cf. Brockett (2012), i.e. copies of one dynamical system, which admit a certain degree of heterogeneity within the ensemble, and which are subject to the restriction that they may only be manipulated and observed as a whole, i.e. through the application of a broadcast signal and through the observation of aggregated population outputs, respectively.

A particularly suitable and relevant example of an ensemble as considered in this thesis is given by cancer cell populations, or more generally, heterogeneous cell populations. A heterogeneous cell population consists of a large number of cells that are genetically identical, but which, even so, admit a certain degree of heterogeneity due to phenotypic variances. Furthermore, it is inherent to the experimental setups that cells within the population cannot be manipulated individually, but only through a common signal, e.g. by a common stimulus through a drug treatment or through the application of light pulses, as recently considered in Milias-Argeitis et al. (2011). Similarly, in practice one typically cannot track the evolution of dynamic processes within individual cells over time, but only the evolution of aggregated quantities of individual cells in the population. These are the aforementioned restrictions, which are simply owed to specific practical limitations that one typically faces when dealing with large populations of systems. The understanding of this critical combination of heterogeneity within the population on the one hand and interaction at the population level only on the other hand is of fundamental importance, and our current lack thereof is drastically illustrated by the extreme resistance of cancer that one is often faced with in current strategies in cancer treatment.

Recent years have witnessed the emergence of several other conceptually similar problems originating from a variety of different applied fields ranging from quantum control (Brockett and Khaneja, 2000; Li and Khaneja, 2006, 2009) to process engineering (Wang and Doyle, 2004) and robotic swarms (Becker and Bretl, 2012; Becker et al., 2014; Kingston and Egerstedt, 2010). These seemingly different problems are indeed intrinsically related by the consideration of populations of nearly identical systems and the manipulation and observation at the population level only; these two premises can be viewed as their common theme.

The ideas related to this theme have already brought forth challenging problems to the field of control theory, and motivated the introduction of new control theoretic concepts. Yet, despite this increasing interest, the current systems theoretic understanding of even the basic aspects of the problems centered around this class of systems remains very limited, and there are only few general results.

In the following section, we proceed with describing the basic and most relevant control theoretic problems related to the consideration of ensembles in more detail.

1.1 Control Theory of Ensembles

Mathematically, an ensemble is typically described in terms of a density function over a state space, cf. Wiener (1938). Therefore, the study of ensembles is closely related to the idea of studying the dynamics of a (probability) density function, rather than the dynamics of a single particle, under the flow of a general dynamical system. This idea has been articulated and considered several years ago within the study of dynamical systems, see e.g. Lasota and Mackey (1994) and references therein. Within the control community, this viewpoint was advocated e.g. in Brockett (2007, 2012). On the one hand, this is motivated by the aforementioned emerging problems in the control of ensembles. On the other hand, the consideration of densities also has a familiar point of contact with classical control theory, such as in the feedback control of uncertain plants, in which probability distributions are used to describe uncertainty. For instance, in Brockett (2007) it is discussed why, for certain situations, it is more suitable to investigate (optimal) control strategies that perform well not only for *one* initial state but rather for a continuous *distribution* of initial states.

Even though both the ensemble and the stochastic control perspective initially are subject to the same mathematical description, a sharp distinction has to be drawn between both problem types when it comes to the interpretation of the models. For example, if a probability distribution is used as a description for uncertainty, measurement data is typically viewed as a realization of the probability distribution, and from a practical point of view as *the* measurement of *the* plant that is subject to uncertainty. If, on the other hand, the probability distribution is used to describe a distribution of particles or cells, measurement data gathered in typical experiments is viewed as a vast number of samples taken from the probability distribution, or, upon further idealization, the whole distribution itself. This may initially be regarded as a rather unfamiliar situation from the perspective of classical stochastic control, which, at the same time, illustrates how the recent considerations of population systems require the consideration of fundamentally different ideas and concepts.

In the following, we discuss the basic control theoretic questions related to ensembles in greater detail. These can be summarized in terms of controllability and observability problems, where the focus of this thesis will be on the latter. The controllability of ensembles has already been studied in several (different but related) contexts, whereas the observability of ensembles has not yet been studied.

Control of the Liouville Equation and Ensemble Controllability

Along with the paradigm shift of considering the evolution of densities rather than single points naturally comes the replacement of the usual model for a nonlinear control system $\dot{x}(t) = f(x(t), u(t))$ by the partial differential equation

$$\frac{\partial}{\partial t} p(t, x) = -\operatorname{div}(p(t, x) f(x, u(t))), \quad p(0, x) = p_0(x).$$

This equation is known as the (controlled) Liouville equation, which is a transport equation that describes how a density $p_0 : \mathbb{R}^n \to \mathbb{R}$ of initial states is advected with the flow of a nonlinear differential equation of the form $\dot{x}(t) = f(x(t), u(t))$.

The density-based viewpoint has already found its way into promising new concepts and applications in control theory over the recent years. For example, Brockett (2012) discusses how one can use the Liouville equation to formulate optimal control problems which aim at maximizing the domain of attraction of a given equilibrium by considering the minimization of cost functionals of the type $J = \int_0^\infty \int_{\mathbb{R}^n} \tanh(\|x\|)\, p(t, x) \,\mathrm{d}x \,\mathrm{d}t$.

An earlier example of an approach employing a very similar idea is given in Rantzer (2001), where a dual to Lyapunov's stability theorem for proving "generic asymptotic stability" of a nonlinear system was introduced. Therein, the key idea is inherently related to the aspects mentioned above, and, furthermore, has been shown to have a direct connection to the Liouville equation (Rajaram et al., 2010).

Another fundamental aspect of the approach employing the Liouville equation is that by considering densities rather than single "particles" as a description of the state of a system, the difference between pure open-loop (broadcast) control and feedback control becomes particularly distinct, cf. Brockett (2012). Therefore it is of great interest to understand these density-based ideas also for the purpose of better understanding the mechanisms of classical feedback. To this end, the general conceptual question of controllability of the Liouville equation asks to which extent one can influence the evolution of a density in the vector field $\dot{x}(t) = f(x(t), u(t))$ through the use of control. The existing results on this problem are mostly summarized in Brockett (2007, 2012), and besides these, only little is known about the controllability of the Liouville equation.

There are, however, many useful results on the controllability of the related equation

$$\frac{\partial}{\partial t} x(t, \theta) = A(\theta) x(t, \theta) + B(\theta) u(t), \quad x(0, \theta) = x_0(\theta) \in \mathbb{R}^n. \tag{1.1}$$

The study of these linear parameter-dependent ensembles is a topic of active research in the field that is now called ensemble control, as systems of this specific type are quite effective at describing many practical ensemble control problems, e.g. in quantum control (Brockett and Khaneja, 2000; Li and Khaneja, 2006, 2009).

A central problem in ensemble control is the controllability of an ensemble (1.1) through open-loop input signals which are, in particular, *independent* of the parameters of the systems. If the parameter set consists of finitely many values, then this corresponds to the classical problem of controlling finitely many linear systems in a parallel connection (Fuhrmann, 1975). The key results on the ensemble controllability of linear parameter-dependent systems can be found in Helmke and Schönlein (2014); Li (2011).

Ensemble Observability

In this section, we discuss the counterpart to the aforementioned paradigm shift for state reconstruction of dynamical systems, which is the focus of this thesis. The consideration of this counterpart is indeed also motivated by practical state estimation problems for populations of dynamical systems that are emerging in a wide range of applied fields. Therein, the crucial point is that only aggregated output measurements of individual systems are available, a circumstance that is perhaps best illustrated by the example of state and parameter estimation for heterogeneous cell populations, cf. Hasenauer et al. (2011a,b). As mentioned in the beginning of the introduction, these populations of genetically identical cells, i.e. structurally identical dynamical systems, admit heterogeneous initial states or parameters. The drastic effect of such heterogeneity is exemplified by cancer cell populations: it is well-known that these cell populations behave highly heterogeneously, which explains observations such as the survival of subpopulations of cancer cells when a cell death stimulus is applied to a cancer cell population. Thus, as a first step in the study of such heterogeneous cancer cell populations, one would like to understand the underlying heterogeneity in initial states and parameters, i.e. their distribution within the population. However, measurement data in the context of cell populations consists mostly of so-called *population snapshots*, which are provided by high-throughput devices such as flow cytometers, (Hasenauer et al., 2011a; Herzenberg et al., 2006). The general concept of population snapshots is illustrated in Figure 1.1.

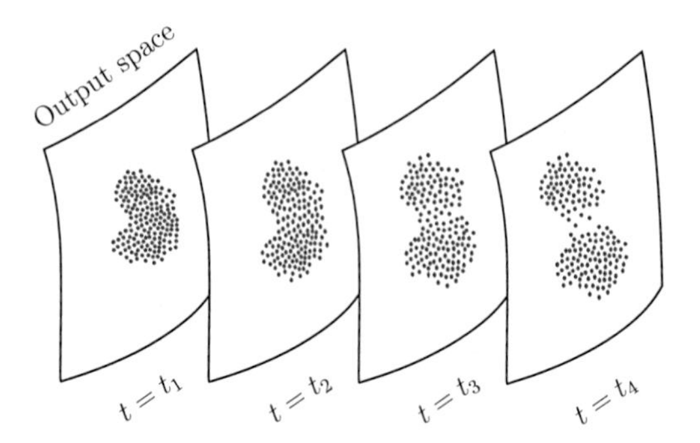

Figure 1.1: Illustration of the concept of population snapshots. In each of the four time steps, a snapshot of certain output values of a population is taken. The crucial point is that in a snapshot, information relating an output value to the individual producing that output value is missing.

A population snapshot, taken at a given instance in time, provides a vast number of output measurements. Yet, at the same time, information relating a measurement to the individual system that produced the measurement is not provided. In fact, heterogeneous cell populations portray a particularly drastic example for state estimation problems of ensembles, since measuring e.g. protein concentrations within a cell often results in killing the cell, making it impossible to measure that cell again. In particular, the measurements given in different snapshots inevitably stem from different cells.

The key to approach this particular state estimation problem for heterogeneous cell populations is to take its particularities into account via a probabilistic framework. Indeed, viewing a population snapshot as a collection of a large number of samples from a probability distribution of outputs yields an adequate first description. Thereby, the justification for this idealization is the large number of cells being considered, so that, in particular, the influence of the invasive measurement on the cell population can be neglected. Similarly, the large number of cells contained in a snapshot can be used to justify the further idealization of the vast number of samples as the output distribution itself. Thus, for the considered state and parameter estimation problems for heterogeneous cell populations, the output distribution is an adequate description for the output of the population. We thus consider the observability problem for a population that is described through a continuous non-parametric probability distribution defined over the states of individual systems, and which is evolving under

$$\begin{aligned} \dot{x}(t) &= f(x(t)), \quad x(0) = x_0, \\ y(t) &= h(x(t)), \end{aligned}$$

where $x_0 \sim \mathbb{P}_0$ is a random vector with a probability distribution $\mathbb{P}_0$. The core of the reconstruction problem can then be formulated as reconstructing the initial distribution of the population from the time evolution of its output distribution.

The estimation of specific distributions in heterogeneous cell populations from population snapshots has been recently considered from a practical point of view (Hasenauer et al., 2011a,b; Zechner et al., 2012). While a wide range of numerical methods for solving such inverse problems is available (see Banks et al. (2012) for a recent review), the underlying estimation problem has not yet been considered from the systems theoretic point of view described above. Although typical ad hoc optimization-based reconstruction methods are designed such that the solutions produced by the methods eventually fit the measurement data well, the actual relation between the solution produced by a specific method and the real distribution in the population is typically not discussed. Moreover, there is no result for guaranteeing the uniqueness of an estimate a priori, despite its fundamental importance to practical estimation problems.

In summary, the study of ensembles in control theory has already led to the introduction of new control theoretic concepts and has already been linked to several important control theoretic problems. Yet, many fundamental problems related to ensembles are still not satisfactorily understood. A major goal within the study of ensembles in control theory is to establish a coherent theoretical foundation for the various emerging applied problems related to ensembles of dynamical systems. In view of the development of modern control theory, it is natural to build such a theory for ensembles around the key concepts of controllability and observability.

1.2 Contributions and Outline of the Thesis

With this thesis we contribute to the basic foundation of a control theory for ensembles of dynamical systems by introducing the concept of ensemble observability of dynamical systems, which can be viewed as a counterpart to the recently considered concept of ensemble controllability. Moreover, we introduce a systems theoretic framework in which this concept is studied and characterized.

Ensemble observability is a system property of a dynamical system with output which characterizes whether or not a continuous distribution of initial states can be reconstructed from observing only the evolution of the resulting output distribution over time. In particular, this concept takes the role of a structural identifiability property for the aforementioned practical state and parameter estimation problems. To complement the study of ensemble observability with respect to continuous distributions, we also consider the discrete counterpart that is obtained by replacing the continuous initial state distribution and the assumption of continuous measurements with discrete ones, respectively. This problem description captures the situation in which one considers a group of finitely many systems, where the premise of population level measurements is reflected in the output measurements of the systems in the group being received at the same time, without any reference between the output measurements and the systems producing them. This discrete version of the ensemble observability problem is related to the problem of multitarget tracking, which, despite its long history in control applications, has not been the subject of a basic theoretical study.

In the following, we provide a more detailed description of the main contributions of the thesis, arranged by its two main chapters.

Chapter 2: Ensemble Observability of Dynamical Systems

The classical question of observability asks whether it is possible to reconstruct the initial state of a finite-dimensional system through the observation of the evolution of an output, as introduced by Kalman (1960), see also Kalman (1963). The concept of observability has become one of the fundamental concepts of modern control theory. In the first part of this thesis, we present a novel yet very natural extension of the notion of observability to the framework of ensemble control. We study the question under which conditions it is possible to reconstruct a non-parametric, continuous probability distribution of initial states when given only knowledge of the evolution of the probability distribution of outputs.

To address the essence of this ensemble observability problem, we first consider the problem in its simplest setting. Given matrices $A \in \mathbb{R}^{n\times n}$ and $C \in \mathbb{R}^{m\times n}$, we consider the linear time-invariant system

$$\begin{aligned} \dot{x}(t) &= Ax(t), \quad x(0) = x_0, \\ y(t) &= Cx(t), \end{aligned}$$

in which x_0 is, however, not considered to be a particular point in $\mathbb{R}^n$, but rather a random vector with a probability distribution $\mathbb{P}_0$ which is unknown.

The ensemble observability problem for linear systems is then to reconstruct the distribution $\mathbb{P}_0$ of initial states given the matrices A and C, as well as the evolution of the resulting distribution of the outputs $y(t) = Cx(t)$.

Our first approach to this problem is in the spirit of inverse problems: we exploit the relation between output and initial state distribution, i.e. that the output distribution is the pushforward measure of the initial state distribution with respect to the forward mapping $x \mapsto Ce^{At}x$. This leads to the discovery of an inherent link between ensemble observability of linear systems and a classical problem in mathematical tomography (Natterer, 1986). Although this appears to be the first result to link the systems theoretic concept of observability to mathematical tomography problems, the connection is in fact very natural: the core of both problems is the inference of some internal state of a system or object from the external measurements given in terms of outputs or radiographs, respectively. By considering the more general observability problem with respect to continuous densities instead of single points in state space, we are able to also establish this connection in mathematical terms.

This novel perspective of the (ensemble) observability problem as a tomography problem is illustrated in several examples and made precise in mathematical terms. We use the available framework of mathematical tomography to formulate characterizations of ensemble observability of linear systems. As in mathematical tomography, the characterizations are formulated in the language of algebraic geometry, which yields a geometric interpretation as a richness condition. In the language of systems and control theory, this may be described as a persistence of excitation condition. This first characterization provides useful insights into the problem and also reveals a gap between classical observability and ensemble observability. Furthermore, we demonstrate that mathematical tomography also provides a suitable framework for addressing the practical reconstruction of an initial state distribution from the measured output distributions.

Our second approach to study ensemble observability of linear systems is in a more systems theoretic spirit: we consider the reconstructability of the moments of the initial state distribution. To this end, we recall the framework of tensor systems (Brockett, 1973), in which the moment dynamics of the state and output distributions can be conveniently described. In particular, the dynamics of moments of the same order are linear and closed. This approach yields a characterization of ensemble observability in terms of the usual observability of the linear (tensor) systems describing the dynamics of the moments of a given order. We furthermore establish a direct connection to our first approach, which may be described as a duality of the two approaches. For the special case of observable single-output systems (A, C) with A having distinct eigenvalues, we provide a reformulation of the resulting characterization of ensemble observability in terms of a verifiable condition on the spectrum of A.

In the last part of this chapter, we consider the ensemble observability problem for nonlinear systems. In the nonlinear case, a geometric perspective on the problem as a nonlinear tomography problem is still valid and insightful. This is illustrated by means of a two-dimensional nonlinear oscillator. For this example, we also showcase the application of the moment-based framework. Even though the moment dynamics do not close as a result of the nonlinearity, a proof of ensemble observability for the specific example is established by exploiting the structure of the particular moment dynamics.

Chapter 3: Sampled Observability of Discrete Linear Ensembles

In the second main part of the thesis, we consider the discrete version of the general ensemble observability problem. Based on our treatment of the case of continuous initial state distributions in the first part of the thesis, it is indeed natural to consider the discrete counterpart which results from replacing both the continuous distribution and the continuous measurement time with discrete ones, respectively. The resulting problem is in fact equivalent to the consideration of a finite set of linear systems

$$\begin{aligned} \dot{x}^{(i)}(t) &= Ax^{(i)}(t), \quad x^{(i)}(0) = x_0^{(i)}, \\ y^{(i)}(t_k) &= Cx^{(i)}(t_k), \end{aligned}$$

with $i \in \{1, \ldots, N\}$ and $k \in \{1, \ldots, M\}$, in a sampled-data theoretic framework. The question is under which conditions the set of initial states $x_0^{(i)}$ can be determined from observing only the discrete distribution of the output at discrete time points. It is imperative to note that, unlike in the framework of the general ensemble observability problem, in the present framework, we may actually speak of individual systems. While these individual systems are dynamically decoupled, a coupling is introduced by the premise that only the distribution of the output can be observed, which corresponds to the fact that at each measurement instance we essentially can only measure the set

$$Y(t_k) := \{y^{(1)}(t_k), \ldots, y^{(N)}(t_k)\}.$$

Thus, while we do obtain all N output measurements, we do not know which output measurement corresponds to which system, i.e. we are lacking the actual associations $y^{(i)}(t_k) \mapsto i$. A typical example that illustrates such a situation is given by social populations, in which, often due to privacy issues, the output measurements of individual systems must be treated as "statistics" without reference to the individual system that produced it. Another way of putting this is that the measurements of the systems are received in an anonymized manner.

It turns out that special cases of this discrete ensemble observability problem have already been considered extensively in the control theory literature since the 1970s under the name of multiple target tracking, or, multitarget tracking, see e.g. Bar-Shalom (1978); Blom and Bloem (2000); Kamen (1992); Smith and Buechler (1975), though from a more practical point of view. Therein, the premise of anonymized output measurements is referred to as a so-called data association problem and is attributed to the use of sensors that do not provide the actual associations. Previous works in multitarget tracking mainly focused on developing approaches for practical solutions. Over the years, several approaches for deriving filters for multitarget tracking have been proposed, the most prominent ones being based on a probabilistic framework, in which the most likely associations are sought (Blom and Bloem, 2000; Chang and Bar-Shalom, 1984), and those approaches based on transforming the received measurements into a set of sums of symmetric products of these (Kamen, 1992). The latter approach completely removes the need for associations between output measurements and individuals, yet one is trading the problem of data association with a polynomial nonlinearity of the newly defined output measurements (Kamen, 1992; Kamen and Sastry, 1993).

As an example, consider the case of three systems with a single-output snapshot given by $(\tilde{y}^{(1)}, \tilde{y}^{(2)}, \tilde{y}^{(3)})$, where $\tilde{y}^{(i)} = y^{(\pi(i))}$ for a permutation π which is undisclosed to us. Since the actual association is not known, the data in the output snapshot cannot be used directly. The idea of symmetric measurement equations (Kamen, 1992) is to encode the data in such a way that no information is lost, and so that the need for data association is removed. A typical symmetric measurement equation is given by the nonlinear encoding

$$\psi(\tilde{y}^{(1)}, \tilde{y}^{(2)}, \tilde{y}^{(3)}) = \begin{pmatrix} \tilde{y}^{(1)} + \tilde{y}^{(2)} + \tilde{y}^{(3)} \\ \tilde{y}^{(1)}\tilde{y}^{(2)} + \tilde{y}^{(1)}\tilde{y}^{(3)} + \tilde{y}^{(2)}\tilde{y}^{(3)} \\ \tilde{y}^{(1)}\tilde{y}^{(2)}\tilde{y}^{(3)} \end{pmatrix}. \tag{1.2}$$

As the name suggests, a key feature is that this transformation is symmetric in the sense that the ordering of the output measurements $\tilde{y}^{(1)}, \tilde{y}^{(2)}, \tilde{y}^{(3)}$ is irrelevant, which is important to obtain a well-defined encoding. Furthermore it can be shown, typically by considering the Jacobian matrix of this transform, that $\psi(\tilde{y}^{(1)}, \tilde{y}^{(2)}, \tilde{y}^{(3)})$ determines the snapshot data $(\tilde{y}^{(1)}, \tilde{y}^{(2)}, \tilde{y}^{(3)})$ uniquely. To solve the multiple target tracking problem practically, one typically employs Extended Kalman Filters (Kamen and Sastry, 1993). Thus, the analysis provided in this chapter may also be viewed as a study of the multi-target tracking problem from a theoretical point of view, which is, again, also motivated by our study of the conceptually more general ensemble observability of linear systems.

We first employ a measure theoretic description for the present setup, similarly to the treatment of the continuous ensemble observability problem. The geometric illustration of this approach eventually leads to geometric and algebraic conditions under which the initial states of the ensemble can be uniquely reconstructed. A first sufficient condition for sampled observability of discrete ensembles can be illustrated in terms of a geometric richness condition. While for the classical sampled observability problem of one linear system the kernels $\ker Ce^{At_k}$ only need to intersect trivially, for the sampled observability of discrete ensembles the kernels $\ker Ce^{At_k}$ need to be even more "scattered in space".

Following the application of a moment-based framework in the general ensemble observability problem in Chapter 2, we introduce a finite analogue of moments for discrete ensembles, which can also be viewed as a generalization of the basic idea of symmetric measurement equations, such as (1.2). These specific moments of discrete ensembles are furthermore shown to be related by tensor systems, just as in the situation for continuous probability distributions. Therefore, by considering the discrete ensemble observability problem, we shed some further light on the connection to the continuous ensemble observability problem studied in Chapter 2.

In the last part of this chapter, we study the sampled observability problem for discrete ensembles in a sampled-data theoretic framework. Through this approach, we are able to provide a series of non-pathological sampling results for different types of discrete ensembles of linear systems, which is of interest in its own right. The approaches for establishing these non-pathological sampling results for discrete ensembles are based on a general non-pathological sampling result for irregularly sampled linear systems, which is briefly discussed. In the simplest case of a discrete ensemble comprised of N linear systems (A, C), the non-pathological sampling result requires a sampling frequency that is N times higher than the critical sampling frequency of the system (A, C).

2 Ensemble Observability of Dynamical Systems

In this chapter, we introduce the notion of ensemble observability of dynamical systems. Ensemble observability is a systems theoretic property of a dynamical system with output, which characterizes a system in terms of whether or not one can reconstruct a continuous non-parametric distribution of initial conditions from observing the evolution of the resulting output distribution over time. One motivation for studying this concept is that it serves as a prototype problem for a class of practical state estimation problems that deal with populations consisting of vast numbers of individual systems. That is, in a typical such practical state estimation problem, one is able to gather vast numbers of output measurements of the systems in the population but can neither directly influence which systems to measure a priori nor trace back a measurement, a posteriori, to the particular system that produced that measurement. More generally, it was discussed in the introduction that an adequate model for describing these circumstances is through viewing the measurements as samples drawn from the output distribution. With the further step to idealize the large number of samples as the output distribution itself we arrive at the ensemble observability problem.

We provide several illustrations of the ensemble observability problem and introduce a theoretical framework in which ensemble observability can be studied from different viewpoints. The different approaches will be further shown to be dual in a sense to be made more precise. We first introduce an approach that is in the spirit of inverse problems, which eventually reveals that the ensemble observability problem is inherently a mathematical tomography problem. The characterization of ensemble observability for a linear system obtained through this connection is described in the language of algebraic geometry. Furthermore, the natural connection to tomography yields a convenient framework for the practical reconstruction problem. We introduce a second approach which is concerned with the observability of moments of the probability distributions, allowing for a more systems theoretic exposition of the ensemble observability problem. Moreover, we provide a comprehensive discussion on the inherent and insightful connection to the former approach. The systems theoretic approach is able to produce characterizations of ensemble observability which are more general and which allow us, for example, to formulate the result that for the special case of single-output systems in which the system matrix has distinct eigenvalues, the system is ensemble observable for the class of moment-determinate distributions if the non-zero eigenvalues of the system matrix are linearly independent over the rational numbers. Towards the end of this chapter, we also discuss the ensemble observability problem for nonlinear systems.

Parts of the results presented in this chapter are based on Zeng and Allgöwer (2015); Zeng et al. (2014, 2016b), as well as Waldherr, Zeng, and Allgöwer (2014).

2.1 Problem Formulation

As discussed in the introduction, recently a number of basic questions in connection with the observability problem for ensembles of dynamical systems have been identified. One such problem is the identifiability of the state distribution that is to be reconstructed from the time evolution of the corresponding output distribution. In order to capture the essence of the observability problem for ensembles, we first consider it in the simplest setting. Given matrices $A \in \mathbb{R}^{n\times n}$ and $C \in \mathbb{R}^{m\times n}$, we consider an observability problem for the linear time-invariant system

$$\begin{aligned} \dot{x}(t) &= Ax(t), \quad x(0) = x_0, \\ y(t) &= Cx(t), \end{aligned} \tag{2.1}$$

where the initial state x_0 is a random vector, i.e. a multivariate random variable, with a continuous probability distribution $\mathbb{P}_0$. Under this assumption, the output $y(t)$, at some time $t \geq 0$, is also a random vector whose distribution is denoted by $\mathbb{P}_{y(t)}$, or, more explicitly, by $\mathbb{P}_{y(t)|\mathbb{P}_0}$ to highlight the dependency on the initial distribution $\mathbb{P}_0$.

We note that our use of probability distributions is to describe the population of structurally identical linear systems as a deterministic particle system and that it is not used as a description of randomness in a stochastic sense; albeit from a strictly mathematical point of view, there is no sharp distinction so far. The difference between the ensemble observability framework and the stochastic control point of view becomes clearer when considering the type of measurements that are available, which are the actual probability distributions $\mathbb{P}_{y(t)}$, as opposed to single realizations of $\mathbb{P}_{y(t)}$.

The resulting notion of observability for such ensembles is then given as follows.

Definition 2.1 (Ensemble observability of linear systems)**.** A linear system (2.1) is said to be *ensemble observable for a class of continuous probability distributions* P, if

$$(\forall t \geq 0 \;\; \mathbb{P}_{y(t)|\mathbb{P}_0'} = \mathbb{P}_{y(t)|\mathbb{P}_0''}) \;\; \Rightarrow \;\; \mathbb{P}_0' = \mathbb{P}_0'',$$

for all distributions $\mathbb{P}_0'$ and $\mathbb{P}_0''$ in P.

In anticipation of the following sections, we note that a linear system cannot be ensemble observable for the class of *all* continuous probability distributions in general. This is why in the definition of ensemble observability of a linear system the class of probability distributions to be considered is explicitly specified. It will turn out that the most general results on the ensemble observability of linear systems are those in which the considered class is that of *moment-determinate* probability distributions.

The remainder of this chapter is devoted to the study of the ensemble observability problem from several different directions. In a first approach we treat the ensemble observability problem as a generic inverse problem in a measure-theoretical framework. This will reveal a close connection to mathematical tomography problems, by which a class of ensemble observable systems can be characterized. The second approach is in a more systems theoretic spirit in which we aim to reconstruct the initial state distribution by means of reconstructing the moments of the initial state distribution. We then establish a direct connection between these two frameworks, yielding the aforementioned duality of the two approaches presented in this chapter.

2.2 Illustration of the Ensemble Observability Problem

We start this section with an example which illustrates a first perspective of the ensemble observability problem. This perspective may be referred to as a "dynamic" one, which appears rather natural from a control theoretic point of view.

Example 2.2. Consider a harmonic oscillator of the form

$$\begin{aligned} \dot{x}(t) &= \begin{pmatrix} 0 & 1 \\ -1 & 0 \end{pmatrix} x(t), \quad x(0) = x_0, \\ y(t) &= \begin{pmatrix} 1 & 0 \end{pmatrix} x(t), \end{aligned} \tag{2.2}$$

as the underlying linear system. Given an initial distribution, which we describe in terms of $x_0 \sim \mathbb{P}_0$, the output measurements, i.e. the output distributions, are obtained from the propagation of $\mathbb{P}_0$ with the flow of the harmonic oscillator which are then marginalized over the second coordinate. This situation is illustrated in Figure 2.1.

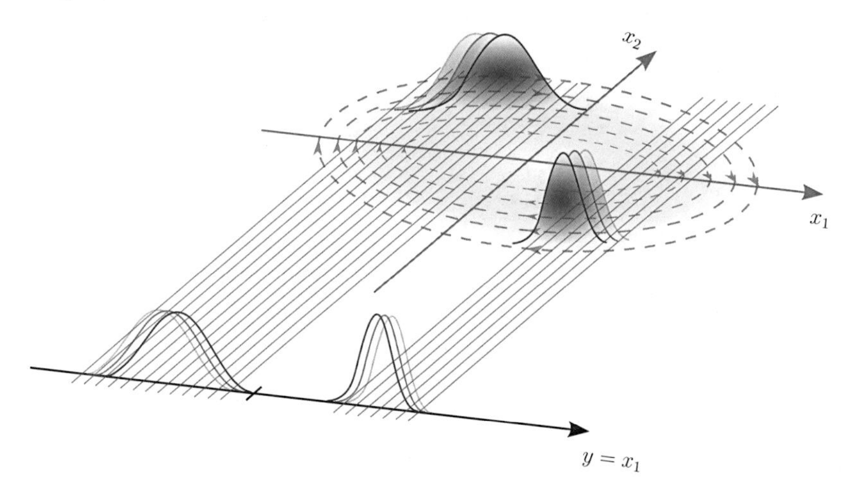

Figure 2.1: Illustration of the dynamic viewpoint of the ensemble observability problem for a two-dimensional harmonic oscillator with a bimodal initial density. The upper right shows the evolution of the initial density. The evolution of the output density is shown in the lower left.

The ensemble observability problem for this particular example is to reconstruct the two-dimensional density shown on the upper right of Figure 2.1 from observing only the time-evolution of the projections shown on the lower left.

The example hints that the study of ensemble observability is a so-called inverse problem, just as the classical controllability and observability problems can be naturally viewed as well, cf. Luenberger (1969).

This suggests to treat this problem from an inverse problems perspective. Before we proceed with addressing the inverse problem, it is reasonable to first discuss the direct problem, i.e. how, in mathematical terms, the output distribution evolves from the initial distribution under the finite-dimensional linear system (2.1). Having established this, we proceed with investigating to which extent we can use our understanding of this direct problem to progress towards the opposite direction for the inverse problem. This will then lead to a geometric illustration of the ensemble observability problem that will be the starting point for a first characterization of ensemble observability of linear systems the framework of mathematical tomography.

Since the output distribution $\mathbb{P}_{y(t)}$ is by definition the probability distribution of the random vector $y(t)$, and since the output is related to the initial state via $y(t) = Ce^{At}x_0$, the distribution $\mathbb{P}_{y(t)}$ is recognized as the *pushforward measure*, or simply *pushforward* of $\mathbb{P}_0$ under the mapping $x \mapsto Ce^{At}x$, which is defined through

$$\mathbb{P}_{y(t)}(B_y) = \mathbb{P}_0((Ce^{At})^{-1}(B_y)) \tag{2.3}$$

for any measurable set $B_y \subset \mathbb{R}^m$. That is, to compute the probability $\mathbb{P}_{y(t)}(B_y)$ for a Borel set $B_y \in \mathcal{B}(\mathbb{R}^m)$, one pulls back B_y via the mapping $x \mapsto Ce^{At}x$ to obtain the preimage $(Ce^{At})^{-1}(B_y) := \{x \in \mathbb{R}^n : Ce^{At}x \in B_y\}$, which is then measured via $\mathbb{P}_0$.

Since we are considering continuous distributions $\mathbb{P}_0$, we may reformulate (2.3) as

$$\mathbb{P}_{y(t)}(B_y) = \int_{(Ce^{At})^{-1}(B_y)} p_0(x)\,\mathrm{d}x, \tag{2.4}$$

where $p_0 : \mathbb{R}^n \to \mathbb{R}$ denotes the probability density function of $\mathbb{P}_0$. This establishes the direct problem which can now be used to address the inverse problem of reconstructing $\mathbb{P}_0$ from $\mathbb{P}_{y(t)}$. From the measured output distributions $\mathbb{P}_{y(t)}$, where $t \geq 0$, we know the value of the right-hand side of (2.4) for all $t \geq 0$ and $B_y \in \mathcal{B}(\mathbb{R}^m)$, which is the integral of the initial density p_0 that we are interested in, over the preimages $(Ce^{At})^{-1}(B_y)$. This description of the inverse problem is illustrated in Figure 2.2.

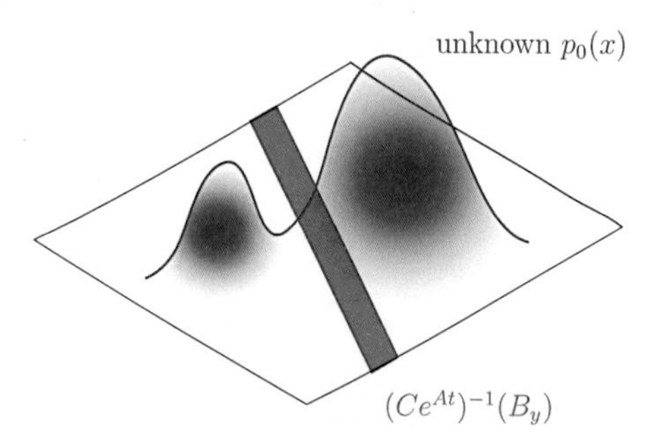

Figure 2.2: The ensemble observability problem formulated as the reconstruction of a density p_0 from its integrals along "strips" $(Ce^{At})^{-1}(B_y)$ for different $t \geq 0$ and $B_y \in \mathcal{B}(\mathbb{R}^m)$.

The main obstruction in reconstructing p_0 from those integrals is the fact that for a fixed $t \geq 0$ and arbitrary Borel sets $B_y \in \mathcal{B}(\mathbb{R}^m)$, information about p_0 in the direction of the "strip" $(Ce^{At})^{-1}(B_y)$ is "integrated out" and thereby lost. Thus, each such piece of information on the initial distribution obtained through the values $\int_{(Ce^{At})^{-1}(B_y)} p_0(x)\,\mathrm{d}x$ does not provide much information about p_0 in general. This is due to the fact that the output matrix $C \in \mathbb{R}^{m\times n}$ does not induce an injective linear mapping in general. However, we may hope that, as time passes, the directions (orientations) of the strips $(Ce^{At})^{-1}(B_y)$ are changing due to the observability properties of (A, C), and that we can eventually infer p_0 by combining the different pieces of partial information. This is the same fundamental problem as in mathematical tomography problems, yielding a direct connection between ensemble observability and mathematical tomography. This connection is perhaps not too surprising from a conceptual point of view, since both problems are well-known to be concerned with inferring an internal density from external projections, which both the radiographs in tomography and the output distributions are.

Before we further study this inverse problem in greater generality, in the following two examples we illustrate and further emphasize the connection between ensemble observability and mathematical tomography problems.

Example 2.3. In this example, we illustrate the viewpoint of the ensemble observability problem as an inverse problem, and more specifically, as a tomography problem. To this end, we reconsider the harmonic oscillator from Example 2.2. We will show that for this particular system the setup of the ensemble observability problem precisely corresponds to the classical prototype problem in computed tomography. Therein the goal is to reconstruct an unknown two-dimensional density, of which one can gather, over time, a full 180° view in terms of projections.

First of all, we recall that instead of the dynamic viewpoint illustrated in the first example, we may as well consider the viewpoint associated with the pushforward relation (2.4). Intuitively, the latter viewpoint corresponds to a situation in which the initial distribution $\mathbb{P}_0$ is not propagated with the flow but held fixed and, instead, the measurement directions are propagated, according to the dynamics of (A, C). This may be regarded as a more "static" perspective, as opposed to the "dynamic" perspective described in the beginning of this section; though it is to be stressed that, in this "static" perspective, the dynamics are not eliminated, but, again, incorporated in the evolution of the directions.

For the considered harmonic oscillator, which rotates clockwise, we have

$$\ker Ce^{At} = \operatorname{span}\left(\left\{\begin{pmatrix} -\sin(t) \\ \cos(t) \end{pmatrix}\right\}\right).$$

Thus the measured output densities in this example can be viewed as projections of the initial density where the angle at which the projections are taken rotates in a uniform counter-clockwise motion. When viewed in this way, the ensemble observability problem for the considered harmonic oscillator is seen to be precisely the prototype example of a two-dimensional tomography problem. The situation is illustrated in Figure 2.3. In the language of tomography, the red lines in Figure 2.3 show the direction of the measurement array which gathers the projections in the respective directions.

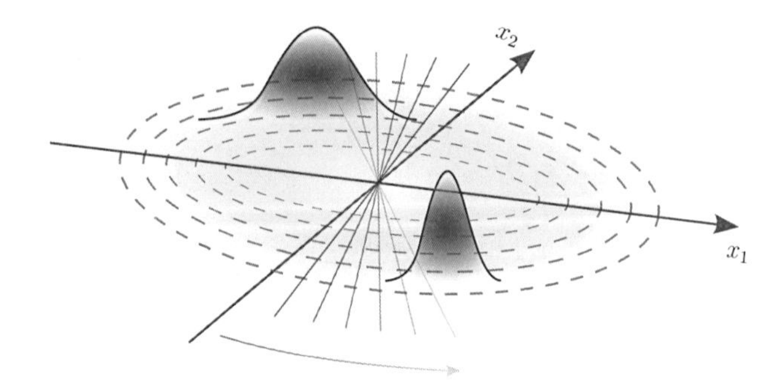

Figure 2.3: This figure illustrates the tomography point of view of the ensemble observability problem for the harmonic oscillator. Therein, one considers the initial density as fixed and incorporates, instead, the dynamics of the system in the rotation of the measurement array which is given by the evolution $t \mapsto Ce^{At}$. The red lines indicate the counter-clockwise rotation of the array.

In the following, we consider a second example which is, perhaps, slightly less specific than the foregoing harmonic oscillator. It is therefore an appropriate example for illustrating a more general situation of the ensemble observability problem.

Example 2.4. Consider the two-dimensional system

$$\dot{x}(t) = \begin{pmatrix} -1 & 1 \\ 0 & 0 \end{pmatrix} x(t), \quad x(0) \sim \mathbb{P}_0. \tag{2.5}$$

We note that this system can be viewed as a very simplistic model of a gene regulatory network described by $\dot{z}(t) = -z(t) + \theta$, with a protein concentration $z(t)$ and some constant parameter θ. In this simple reaction network, the protein is subject to linear degradation and is constantly activated according to the rate θ. The phase portrait of the linear system is illustrated in Figure 2.4.

Furthermore, we consider the two different output matrices

$$C' = \begin{pmatrix} 0 & 1 \end{pmatrix} \quad \text{and} \quad C'' = \begin{pmatrix} 1 & 0 \end{pmatrix},$$

with the first output matrix C' leading to (A, C') being unobservable, and the second output matrix C'' leading to (A, C'') being observable. By considering the two cases of an unobservable and observable underlying system for the ensemble we highlight, from a geometrical point of view, the relevance of the classical observability properties of the underlying linear system for the ensemble observability problem.

By the formulation of the ensemble observability problem in terms of (2.4), it is clear that the properties of the kernels $\ker Ce^{At}$, for a given output matrix C, are crucial for the reconstructability of p_0. In studying the evolution of $\ker Ce^{At}$ it is also helpful to recall the basic relation

$$\ker Ce^{At} = e^{-At}(\ker C).$$

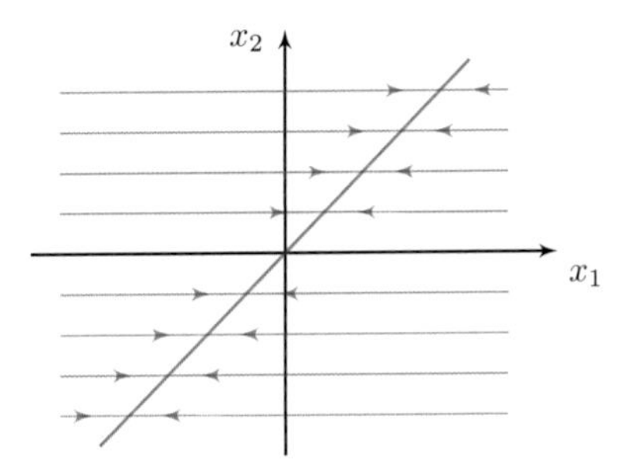

Figure 2.4: The phase portrait of the system $\dot{x}(t) = Ax(t)$ given in (2.5).

Thus, the relevant measurement directions that are indicated by $\ker Ce^{At}$ can be viewed as being obtained from transporting $\ker C$ with the flow of the system $\dot{x}(t) = Ax(t)$ backwards in time, cf. the situation in Example 2.3.

Since $\ker C'$ is precisely the x_1-axis, which is invariant under the flow of the system, the sets $(C'e^{At})^{-1}(B_y)$, for Borel sets $B_y \subset \mathbb{R}$, may be described as "horizontal strips". By only having at hand integrals of p_0 over strips that have the same "orientation", however, one cannot expect to be able to uniquely reconstruct p_0; for example, an arbitrary p_0 shifted along the x_1-axis leaves the resulting output distributions invariant. More generally, the idea is that due to non-observability, one ends up with a non-trivial intersection

$$\bigcap_{t \geq 0} \ker Ce^{At} = \{x_0 \in \mathbb{R}^n : Ce^{At}x_0 \equiv 0\}. \tag{2.6}$$

As before, an arbitrary density p_0 may be shifted along a non-zero vector taken from this intersection, yielding two different densities that yield the same value when integrated over preimages $(Ce^{At})^{-1}(B_y)$, or, in other words, two different but indistinguishable densities. Since this argument holds for an arbitrary unobservable linear system (A, C), this shows that the classical observability of (A, C) is a necessary condition for the ensemble observability of (A, C) for the class of continuous initial distributions.

It is now interesting to see what happens for the observable system (A, C''). There, we find that $\ker C''$ is precisely the x_2-axis. By inspecting the evolution of $e^{-At}(\ker C'')$ in Figure 2.5 we see that the kernels are now tilted by the flow in a counter-clockwise rotation, which is in accordance with the intersection (2.6) being trivial due to observability of (A, C''). Thus, by considering measurements of the output distribution $\mathbb{P}_{y(t)}$ at different points in time, we now obtain integrals of p_0 along strips at different angles, thus gaining more information about the initial density p_0 than in the unobservable case. Yet, it is noted that the available range of angles is still constrained by the observability properties of the system, which will be referred to as a "limited direction" situation. This raises the intriguing question as to whether or not these different pieces of partial information can in fact be put together so as to get full information about p_0.

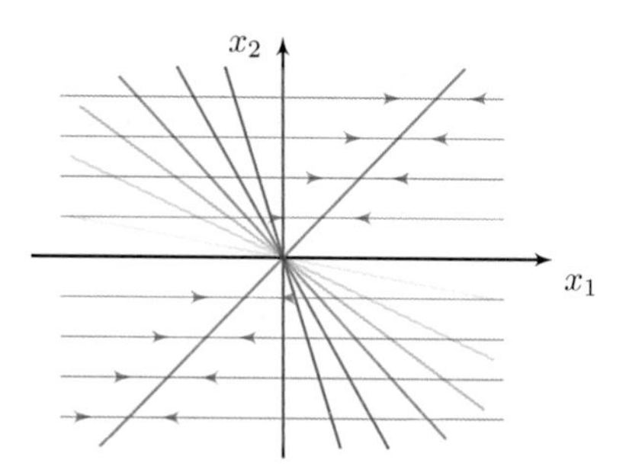

Figure 2.5: The evolution of the kernel $\ker C''e^{At}$ is given by transporting the kernel $\ker C''$ with the flow of $\dot{x}(t) = Ax(t)$ *backwards in time.* This results in a tilting of the kernel $\ker C''$, yielding a trivial intersection of $\ker C''e^{At}$ in accordance with the observability of (A, C''). The color intensities of the kernels indicate advancement of the transported kernels in time.

To further emphasize the connection between the ensemble observability problem and mathematical tomography, we note that in the case of linear systems, we can further reformulate the ensemble observability problem in terms of probability density functions by using the *coarea formula.* The coarea formula relates the integral of a density p_0 over an open set in $\mathbb{R}^n$ to integrals over level sets of a function $H : \mathbb{R}^n \to \mathbb{R}^m$ with $m < n$, i.e.

$$\int_{H^{-1}(B_y)} p_0(x)|J_m H(x)| \,\mathrm{d}x = \int_{B_y} \left(\int_{H^{-1}(\{y\})} p_0(x) \,\mathrm{d}S \right) \mathrm{d}y$$

for any measurable set $B_y \subset \mathbb{R}^m$, where $J_m H$ denotes the m-dimensional Jacobian of $H : \mathbb{R}^n \to \mathbb{R}^m$. With $H(x) = Ce^{At}x$ for the considered linear case, we obtain

$$|J_m H(x)| = \sqrt{\det(Ce^{At}(Ce^{At})^\top)},$$

which is independent of the state $x \in \mathbb{R}^n$. Therefore, in the linear case, we can leverage the coarea formula to eventually obtain an explicit relationship between the output density $p_{y(t)}$ and the initial density p_0, given by

$$p_{y(t)}(y) = \frac{1}{\sqrt{\det(Ce^{At}(Ce^{At})^\top)}} \int_{(Ce^{At})^{-1}(\{y\})} p_0(x) \,\mathrm{d}S. \tag{2.7}$$

More precisely, this is because integrating (2.7) over *any* measurable set $B_y \subset \mathbb{R}^m$ yields the same value as integrating p_0 over the set $(Ce^{At})^{-1}(B_y)$ by virtue of the coarea formula. By the basic pushforward relation (2.3) and the definition of probability density functions, $p_{y(t)}$ is indeed the probability density function of $\mathbb{P}_{y(t)}$. Roughly speaking, passing from the integral over "strips" (2.4) to the surface integral (2.7) can be thought of as a concentration of the information $\int_{(Ce^{At})^{-1}(B_y)} p_0(x) \,\mathrm{d}x$ about p_0 by taking the "width" of a "strip", as illustrated in Figure 2.2, to zero.

In this form, the ensemble observability problem is given in the most classical formulation of a mathematical tomography problem; the problem of reconstructing a density from its integrals over affine subspaces. Such problems of reconstructing a density from integrals over affine subspaces, or more generally manifolds, are one of the hallmarks of the theory of mathematical tomography.

2.3 Characterization via Mathematical Tomography

In the previous section, we demonstrated how the ensemble observability problem is inherently related to problems in mathematical tomography. In this section, we first provide a brief overview of the theory of mathematical tomography and then introduce the mathematical framework. Given this framework, we proceed towards establishing a first solution to the ensemble observability problem given in terms of an analytic characterization of ensemble observability of linear systems.

Background on Mathematical Tomography

Classical tomography may be described as a way to determine the internal structure of an object without having to open it up. Perhaps the best known example for a tomographic problem is computed tomography, which is used for providing cross-sections of e.g. a part of a body for medical diagnosis. Computed tomography is based on the physical properties of an X-ray beam passing through an object, by which properties of the object can be inferred. More precisely, consider an X-ray beam L passing through an object of interest with density $f : \mathbb{R}^2 \to \mathbb{R}$. Let L be parameterized by the variable z. Then the intensity I along L is attenuated according to the Beer-Lambert law (Markoe, 2006)

$$\frac{d}{dz} I(z) = -f(L(z))I(z).$$

By virtue of this law, measuring the intensity

$$I_1 = I_0 \exp\left(-\int_L f(x)\,\mathrm{d}S\right)$$

of the beam L after it went through the object of interest and comparing it with the intensity I_0 at which it was emitted, we can compute the value

$$\int_L f(x)\,\mathrm{d}S = \log\left(\frac{I_0}{I_1}\right).$$

A. M. Cormack, one of the inventors of computed tomography, pursued this problem of reconstructing a density from its line integrals (Cormack, 1963, 1964) and proposed a practical reconstruction method, for which he was awarded the Nobel prize in medicine and physiology in 1969, jointly with G. N. Hounsfield. Only later was it discovered that the mathematical problem has been solved already 50 years earlier by the mathematician J. Radon (Radon, 1917), though for purely mathematical interests.

The general problem of tomography is the reconstruction of a function from its *Radon transform* (Markoe, 2006), which, in its classic form, is a transformation that maps a two-dimensional scalar function f to the transform Rf that is defined on lines L, i.e. $Rf(L) = \int_L f(x)\,\mathrm{d}S$. In the more general n-dimensional case, the Radon transform maps an integrable function $f \in L^1(\mathbb{R}^n, \mathbb{R})$ to its transform $Rf : \mathbb{S}^{n-1} \times \mathbb{R} \to \mathbb{R}$ which is defined by

$$Rf(\omega, p) = \int_{\{x \in \mathbb{R}^n \,:\, \langle \omega, x \rangle = p\}} f(x)\,\mathrm{d}S,$$

whenever the integral exists, cf. Markoe (2006). Moreover, for a unit vector $\omega \in \mathbb{S}^{n-1}$, the function $R_\omega f$ defined by $(R_\omega f)(p) = Rf(\omega, p)$ is called Radon projection along $\omega^\perp$.

We note that the definition of the n-dimensional Radon transform given here can also be generalized to cases in which the integration is taken over affine planes of arbitrary dimension. However, since our analysis will not be based on the classical tomography approach but instead a slightly different, more convenient probabilistic variant of it, we choose to limit ourselves to the most basic case for this brief review. We shall once again note that the immediate connection to the ensemble observability problem is expressed by (2.7), and that we may thereby describe the output densities as some kind of Radon projections of the initial density along $\ker Ce^{At}$.

The solution to the classical tomography problem, i.e. the theoretical invertibility of the Radon transform, is based on a close connection of the Radon transform to the Fourier transform. With the definition of the n-dimensional Fourier transform as

$$(\mathcal{F}_n f)(\xi) = \int_{\mathbb{R}^n} f(x) e^{-i\langle x, \xi \rangle}\,\mathrm{d}x,$$

this connection is given as follows, cf. Markoe (2006).

Theorem 2.5 (Projection-slice theorem). *Consider $f \in L^1(\mathbb{R}^n, \mathbb{R})$ and let $\omega \in \mathbb{S}^{n-1}$ be a unit vector. Then one has the identity*

$$(\mathcal{F}_1 R_\omega f)(\sigma) = \mathcal{F}_n f(\sigma\omega).$$

That is, the one-dimensional Fourier transform of the Radon projection along $\omega^\perp$ is equal to the n-dimensional Fourier transform of the density restricted to the "slice" parameterized in terms of $\sigma\omega$ with $\sigma \in \mathbb{R}$. Therefore, if we have for an integrable function the Radon projections for all "directions" ω, then we know the n-dimensional Fourier transform of f completely. Therefore we know f and have solved in the simplest case the problem of reconstructing a function from all its Radon projections. This is the classical solution to the tomography problem.

Unfortunately, this result does not apply directly to our problem. In contrast to the problem in computed tomography, we may not freely choose the directions at which we can gather Radon projections, but the directions $\ker Ce^{At}$ are inherently determined by the dynamical component of the ensemble observability problem, or, more concretely, the observability properties of the linear system (A, C), as was illustrated in Example 2.4. This issue will be addressed in the next subsection, eventually yielding a first sufficient condition for ensemble observability of linear systems.

Sufficient Conditions for Ensemble Observability

To deal with the general case of subspaces of arbitrary dimension, as well as the limited direction problem, we turn to a more convenient probabilistic description of the mathematical tomography problem. This allows for an elementary treatment that avoids unnecessary technicalities.

First of all, we reformulate the projection-slice theorem in the framework of probability theory. This probabilistic analogue is known as the Cramér-Wold device in probability theory (Cramér and Wold, 1936).

Theorem 2.6 (Cramér-Wold theorem). *A distribution of a random vector X in $\mathbb{R}^n$ is uniquely determined by the family of its pushforwards under the linear functionals $x \mapsto \langle v, x\rangle$ with $v \in \mathbb{S}^{n-1}$.*

Proof. The first step is to relate the characteristic function of the distributions of $\langle v, X\rangle$ to that of X via the simple calculation

$$\varphi_{v_1X_1+\cdots+v_nX_n}(s) = \mathbb{E}\big[e^{is(v_1X_1+\cdots+v_nX_n)}\big] = \mathbb{E}\big[e^{i\langle sv, X\rangle}\big] = \varphi_X(sv). \tag{2.8}$$

Since the left-hand side is given for all $v \in \mathbb{S}^{n-1}$ and all $s \in \mathbb{R}$, by the above identity we know the characteristic function φ_X, and thus the distribution of X. □

To see that the Cramér-Wold theorem is in fact a probabilistic analogue of the projection-slice theorem, we observe that the left-hand side of (2.8) is simply the one-dimensional Fourier transform (modulo the substitution $i \mapsto -i$) of the density of $v_1X_1 + \cdots + v_nX_n$, whereas the characteristic function on the right-hand side of (2.8) is the n-dimensional Fourier transform of the joint density, i.e. the density of X. The density of the random variable $v_1X_1 + \cdots + v_nX_n$, on the other hand, is nothing but the Radon projection of the joint density along $v^\perp$.

It is interesting to note that before the connection between tomography problems and its probabilistic counterpart was pointed out in Rényi (1952), cf. Markoe (2006), the developments in both fields took place independently. The Cramér-Wold device is used in probability theory mostly as a conceptional tool, to be more precise, it is used as a means to reduce a high-dimensional problem to a one-dimensional problem to which one can then apply well-established results. Our use of this result at this point is slightly different as we also exploit its analogy to tomography problems explicitly to address the inverse problem of reconstructing the initial density from the output density.

For the ensemble observability problem we compute the characteristic function of the output distribution to find the relation

$$\varphi_{Ce^{At}x_0}(s) = \mathbb{E}\big[e^{i\langle s, Ce^{At}x_0\rangle}\big] = \mathbb{E}\big[e^{i\langle (Ce^{At})^\top s, x_0\rangle}\big] = \varphi_{x_0}((Ce^{At})^\top s). \tag{2.9}$$

That is, the output distributions yield information about the characteristic function of the initial state distribution on the subspaces

$$\operatorname{im}(Ce^{At})^\top = (\ker Ce^{At})^\perp.$$

This simple insight will be key in formulating characterizations for the uniqueness of reconstruction of the initial density p_0.

First note that by (2.9) we cannot, in general, gather information about the whole characteristic function. This is exactly where we need to draw on analyticity properties of the characteristic function. In mathematical tomography, analyticity of the Fourier transform is typically guaranteed by the standard assumption of bounded support of the considered densities. The assumption of bounded support, however, excludes e.g. Gaussian distributions, which would be unfortunate, at least from a theoretical point of view. We will show, however, that the assumption of bounded support can be relaxed to the following more general assumption on the characteristic function φ_{x_0}. We assume that the mappings $s \mapsto \varphi_{x_0}(sv) = \varphi_{\langle v,x_0\rangle}(s)$, for all non-zero $v \in \mathbb{R}^n$, are real analytic, i.e. can be locally written as a power series about every point in $\mathbb{R}$. The role of this assumption will be further illuminated in a moment-based approach that will be introduced in the next section.

We begin by formulating our main result of this section, which gives a first sufficient condition for ensemble observability with respect to a specific class of initial distributions. Our result follows a uniqueness result for the tomographic reconstruction problem, cf. Theorem 5.2 in Keinert (1989) and Theorem 3.142 in Markoe (2006), which give the most relaxed characterization known in the mathematical tomography literature.

Theorem 2.7. *A linear system (A, C) is ensemble observable for the class of initial distributions for which $s \mapsto \varphi_{x_0}(sv)$, for all non-zero $v \in \mathbb{R}^n$, is real analytic, if*

$$\bigcup_{t\geq 0}(\ker Ce^{At})^{\perp} = \bigcup_{t\geq 0} \operatorname{im}(Ce^{At})^{\top} \tag{2.10}$$

is not contained in a proper algebraic subvariety of $\mathbb{R}^n$.

Thus, a sufficient condition is that the directions generated by $t \mapsto Ce^{At}$ are sufficiently "rich" in the sense that we cannot find a proper algebraic variety in which the union (2.10) is contained. Recall that an algebraic variety of $\mathbb{R}^n$ is the zero set of a polynomial, and that it is proper if it is not $\mathbb{R}^n$. More abstractly, we may rephrase the condition in Theorem 2.7 as the property that the set of "measurement orientations" available through the mapping $t \mapsto Ce^{At}$ form an open (and dense) set in the Zariski topology; see e.g. Shafarevich and Hirsch (1977).

An intuitive description of this richness property in the single-output case is that the curve (or signal for that matter) $t \mapsto Ce^{At}$, which is the normal vector of the relevant hyperplanes $\ker Ce^{At}$, exhibits sufficient "complexity" so that its evolution cannot, in particular, be "captured" in terms of (or "trapped in") the zero set of a non-zero polynomial. For the observability of (A, C) in the classical case, on the other hand, it is merely required that $t \mapsto Ce^{At}$ is not contained in any proper linear subspace, which may also be described as the zero set of some homogeneous polynomial of degree one.

Proof of Theorem 2.7. We show that under the analyticity condition on φ_{x_0} and the assumption that the union (2.10) is not contained in a proper algebraic variety, knowing the characteristic function solely on (2.10) is sufficient to know the characteristic function everywhere.

To this end, we consider two characteristic functions $\varphi_{x_0'}$ and $\varphi_{x_0''}$ such that their difference $h := \varphi_{x_0'} - \varphi_{x_0''}$ vanishes on the union (2.10), i.e.

$$h(\xi) = 0 \;\text{ for all }\; \xi \in \bigcup_{t \geq 0} \operatorname{im}(Ce^{At})^\top. \tag{2.11}$$

By analyticity, we can write for any non-zero $\xi \in \mathbb{R}^n$ and any sufficiently small λ,

$$h(\lambda\xi) = \sum_{p=0}^{\infty} \lambda^p a_p(\xi), \tag{2.12}$$

where the coefficients $a_p(\xi)$ of the power series are given by

$$a_p(\xi) = \frac{i^p}{p!} \left(\mathbb{E}[\langle \xi, x_0' \rangle^p] - \mathbb{E}[\langle \xi, x_0'' \rangle^p]\right),$$

as will be shown in Section 2.4. In particular, a_p is a *homogeneous polynomial.*

Considering (2.11) and (2.12), as well as homogeneity of a_p, we find that the union (2.10) is contained in the algebraic varieties defined by a_p. By the assumption that the union (2.10) is not contained in a *proper* algebraic variety, all polynomials must be trivial, i.e. $a_p \equiv 0$. Since for all non-zero $\xi \in \mathbb{R}^n$ the mapping $\lambda \mapsto h(\lambda\xi)$ is real analytic in a neighborhood of any point of the real axis, $\lambda \mapsto h(\lambda\xi)$ is completely determined by its power series about the origin, which is zero. Therefore we conclude that $h \equiv 0$, i.e. $\varphi_{x_0'} = \varphi_{x_0''}$, and thus $\mathbb{P}_0' = \mathbb{P}_0''$ for the corresponding initial distributions. □

We next derive a sufficient condition for the special case which occurs when the affine subspaces that one is integrating over are one-dimensional. From the perspective of tomography, this case arises in the study of the X-ray transform and can be nowadays considered classic (Markoe, 2006). We note, though, that in view of the ensemble observability problem, the assumptions are clearly rather restrictive.

Proposition 2.8. *If* (A, C) *is observable, and* $\operatorname{rank} C = n - 1$, *then the union* (2.10) *is not contained in a proper algebraic variety.*

Proof. With rank $C = n - 1$, the dimension of $(\ker Ce^{At})^\perp$ is also $n - 1$. Due to the observability of (A, C), the intersection (2.6) is trivial and thus $(\ker Ce^{At})^\perp$, with $t \geq 0$, constitutes an infinite family of pairwise distinct hyperplanes. More precisely, we show in the following that for an observable system (A, C), it cannot happen that

$$\forall t \geq 0 \;\exists i = 1, 2, \ldots \quad \ker Ce^{At} = \operatorname{span}(\{v_i\})$$

for arbitrary countable non-zero vectors $v_i \in \mathbb{R}^n$. First of all, since rank $C = n - 1$, the fact that $\ker Ce^{At} = \operatorname{span}(\{v_i\})$ is equivalent to $Ce^{At}v_i = 0$. Thus, with the definition

$$T_i := \{t \geq 0 : Ce^{At}v_i = 0\},$$

we would need $\bigcup_{i=1,2,\ldots} T_i = [0, \infty)$. But this is impossible since the sets T_i consist of isolated points due to observability of (A, C).

The last step is to recall that an infinite family of distinct hyperplanes cannot be contained in a proper algebraic variety. □

Proposition 2.8 has the following remarkable corollary in the special case of $n = 2$.

Corollary 2.9. *For an observable two-dimensional system* (A, C)*, the union* (2.10) *is not contained in a proper algebraic variety.*

Thus, in the case of $n = 2$ the richness property (2.10) in Theorem 2.7 is satisfied via observability of (A, C) alone. One question that arises at this point is whether or not in general an observable system (A, C) may already generate "directions" rich enough such that the union (2.10) is not contained in an algebraic subvariety. This question is, however, quickly answered to the negative through the following counterexample.

Example 2.10. We shall give an example of an observable linear system (A, C) which generates directions $t \mapsto Ce^{At}$ that are contained in a proper algebraic variety. First of all, we note that Corollary 2.9 implies that in order to find a system that is observable, but for which the union (2.10) is contained in a proper algebraic variety, we need to consider systems with at least three state variables. Consider the system

$$\begin{aligned} \dot{x}(t) &= \begin{pmatrix} 0 & & \\ & -1 & \\ & & -2 \end{pmatrix} x(t), \\ y(t) &= \begin{pmatrix} 1 & 1 & 1 \end{pmatrix} x(t), \end{aligned} \tag{2.13}$$

which is easily seen to be observable in the classical sense, since the diagonal entries are pairwise distinct and every entry in the output matrix is non-zero.

It is straightforward to compute

$$Ce^{At} = \begin{pmatrix} 1 & e^{-t} & e^{-2t} \end{pmatrix},$$

from which it can be seen that the algebraic variety given by the homogeneous polynomial equation

$$x_1 x_3 = x_2^2 \tag{2.14}$$

contains the union (2.10), thus violating the richness condition in Theorem 2.7.

We note that, since Theorem 2.7 only provides a sufficient condition, we are not yet in the position to conclude that system (2.13) is not ensemble observable at this point. We will, however, come back to this example later in the next section. Therein, we first introduce the moment-based approach for the study of ensemble observability, and based on the insights gained there, actually construct distinct, indistinguishable Gaussian initial distributions for the considered system (2.13). This will also shed more light on the relevance of the algebraic geometric aspects that we have discussed so far. Before that, in the next subsection, we show how the well-developed computational reconstruction methods from computed tomography can be used for the practical reconstruction of initial state distributions.

Practical Reconstruction Based on Tomography Methods

In this subsection, we briefly discuss and illustrate how, by virtue of the connection to mathematical tomography, the ensemble observability problem also becomes amenable to computational solutions. We demonstrate that the framework of tomography is also well suited for the practical reconstruction of an unknown initial distribution, and, in particular, we will describe how one can obtain a computational solution to the ensemble observability problem in Example 2.4.

For the reconstruction of the initial density, we can leverage the connection to mathematical tomography described in this section, e.g. by employing a well-known reconstruction technique for solving computed tomography problems, that has come to be known as Algebraic Reconstruction Techniques (Gordon et al., 1970). These reconstruction techniques are based on the basic idea of discretizing the state space into pixels so that the unknown distribution is expressed as a piecewise constant function: for a given discretization in terms of N pixels S_i, the initial density is approximated via

$$p_0(x) \approx \sum_{i=1}^{N} p_i \mathbb{1}_{S_i}(x),$$

where p_i is the value of the pixel S_i, and $\mathbb{1}_{S_i}$ denotes its indicator function. Similarly, a grid is introduced on the output space. In fact, in practical experiments we typically expect that the measured output distributions are given in terms of histograms, which yield an approximation of $\mathbb{P}_{y(t)}(B_y)$ for the different "pixels" $B_y \in \mathcal{B}(\mathbb{R}^m)$ that may be referred to as the "bins" of the histogram in this more practical context. The justification of this approximation is only valid provided that the number of output samples in a given snapshot is sufficiently large.

Given the approximation of p_0 in terms of piecewise constant functions, the integrals $\int_{(Ce^{At})^{-1}(B_y)} p_0(x)\,\mathrm{d}x$ can be approximated by weighted sums of the values of the pixels that the strip passes through. Therein, the weights incorporate the area that the considered strip occupies within the ith pixel. Combining these two approximations, the pushforward relation between $\mathbb{P}_0$ and $\mathbb{P}_{y(t)}$ is eventually expressed in terms of the system of linear equations

$$\sum_{i=1}^{N} \frac{|S_i \cap (Ce^{At})^{-1}(B_y)|}{|S_i|} p_i = \mathbb{P}_{y(t)}(B_y),$$

where $|E|$ denotes the Lebesgue measure of a set $E \subset \mathbb{R}^n$, and both the measurement times t and the "bins" B_y are varied. In this system of linear equations, the unknowns to be solved for are the values p_i of each pixel S_i. More concretely, for setting up the system of linear equations, only those strips $(Ce^{At})^{-1}(B_y)$ in which the "bins" B_y hold a non-zero number of samples are considered; every pixel S_i for which there is a measurement time at which no such strip passes through is necessarily zero.

In order to achieve a certain quality for the reconstruction, one aims at obtaining projections of the unknown density at sufficiently many directions; the time points at which output snapshots are measured need to be chosen accordingly. The resulting system of linear equations is typically very large, but due to the structure of the problem, the underlying matrix is rather sparse.

Iterative projection-based methods such as e.g. Kaczmarz method (Kaczmarz, 1937) can be readily employed for solving such systems of linear equations. The idea of the Kaczmarz method is to start with some initialization $p^{(0)}$ which is then iteratively projected on the affine hyperplanes defined by the linear equation of a single row of the large system of linear equations. The iteration over all rows is, in turn, typically iterated several times itself.

More specifically, for the pixels S_i that do not intersect any transported non-empty bin for a fixed time we can directly set $p_i = 0$ after every iteration, instead of computing the weights for the transported empty bins. Likewise, we set those values $p_i \leq 0$ to zero in every step to ensure non-negative values. As we can see, the approach using Algebraic Reconstruction Techniques is not only very accessible, but facilitates the incorporation of a priori knowledge about the initial density p_0 in the iteration. This, and the fact that we are taking the internal structure of the problem explicitly into account, leads to a well-suited and accurate reconstruction method, which is perhaps also more comprehensible than existing approaches.

Example 2.11. To demonstrate this practical reconstruction scheme on a concrete example, we reconsider the observable linear system that we introduced in Example 2.4. Suppose that the initial density is given by the bimodal density $p_0 = 0.7p_1 + 0.3p_2$ where p_1 and p_2 are densities of normal distributions with means and covariance matrices

$$\mu_1 = \begin{pmatrix} 1 \\ 2 \end{pmatrix}, \quad \Sigma_1 = 0.3^2 \begin{pmatrix} 1 & \\ & 1 \end{pmatrix}, \quad \mu_2 = \begin{pmatrix} 2 \\ 1 \end{pmatrix}, \quad \Sigma_2 = 0.2^2 \begin{pmatrix} 1 & \\ & 1 \end{pmatrix},$$

respectively. For this system, we chose the measurement time points according to our knowledge of the system in terms of the directions of $\ker Ce^{At}$ illustrated in Figure 2.5; in general, the time points should be chosen so as to get a uniform angle distribution. At each time point, we collected 10^5 samples of the output distribution. We recall that in the case of practical reconstruction problems where the data is not given in terms of distributions of outputs but rather in terms of samples of the output distribution, the values $\mathbb{P}_{y(t)}(B_y)$ are approximated via sufficiently accurate histograms of the measured samples. We then constructed the system of linear equations by the aforementioned scheme and applied the Kaczmarz method to solve for the unknown pixel values p_i. The result obtained by this procedure is shown in Figure 2.6 for different numbers of iteration. For smaller numbers of overall iterations, we witness a well-known "distortion" effect which is due to the limited direction situation imposed by the dynamics of the underlying linear system, cf. Figure 2.5 in Example 2.4. This undesired effect can be attenuated by increasing the number of iterations. Yet, a drawback of increasing the number of iterations is the numerical noise that is introduced into the problem.

It is interesting to note that while Theorem 2.7 explicitly requires measurement data of the output densities for infinitely many time points, in practice one can of course only use a finite number of measured output densities. In other words, the theoretical results do not apply directly. To make matters worse, one can even show that there are infinitely many different densities that produce a given finite set of Radon projections exactly, see e.g. Helgason (2011); Markoe (2006). Nevertheless, for practical purposes, we can typically expect the practical reconstruction to yield a small estimation error for sufficiently many measured output densities.

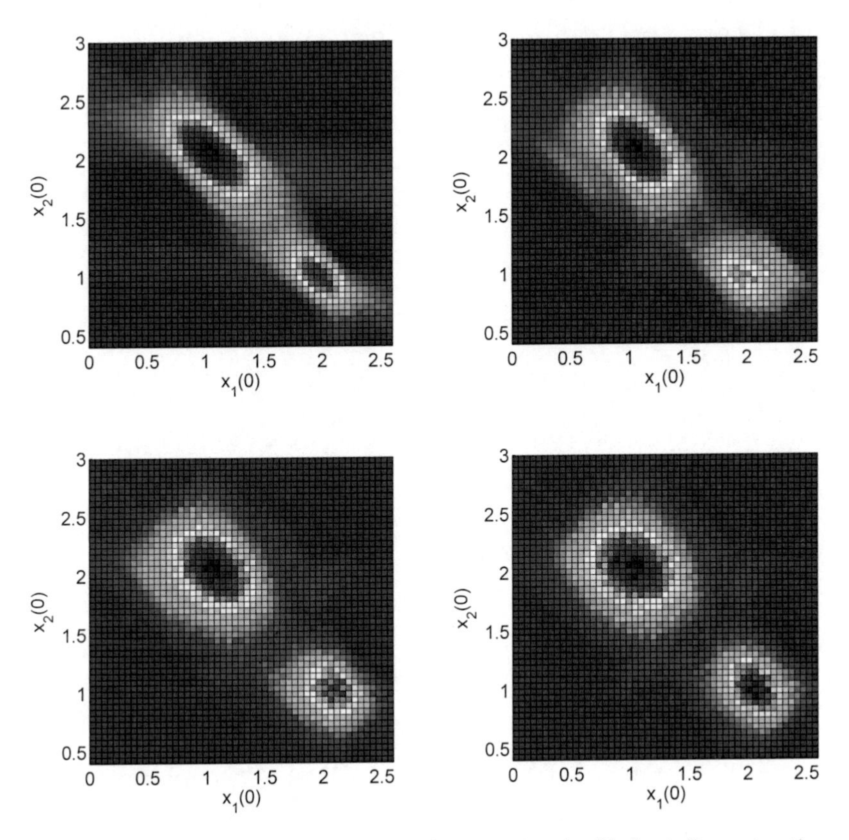

Figure 2.6: Reconstructions of the bimodal density, using the Algebraic Reconstruction Technique with different iteration numbers 1 and 3 (top, from left to right), and 5 and 7 (bottom, from left to right). For a smaller number of iterations, a "distortion" effect is witnessed, which may be described as the result of shearing the original density in a horizontal direction. This is known to be caused by the limited direction situation. This effect is reduced by increasing the number of iterations, as seen in the bottom right plot.

In conclusion, while the established methods from tomography seem to be generally well suited for the practical reconstruction of a distribution of initial states, there are still specific challenges stemming from the dynamic origin of this problem. Moreover, due to the discretization of the state space, the scope of this reconstruction technique is limited to low-dimensional problems. An exhaustive discussion of different tomographic methods, as well as other different reconstruction methods for the ensemble state reconstruction problem is, however, beyond the scope of this thesis.

2.4 Characterization in a Moment-Based Framework

In this section, we present an alternative, systems theoretic approach for characterizing ensemble observability of linear systems. This systems theoretic approach will lead to a description of ensemble observability of linear systems in terms of the classical observability of an infinite family of tensor systems related to the original linear system. Besides providing another perspective on the ensemble observability problem, one advantage of this description is that it allows for a reformulation for a specific class of linear systems in terms of a verifiable condition that only depends on the matrices A and C. Moreover, we will show that the approach presented in this section is inherently related to the approach of the previous section, which is based on the ideas and the language of mathematical tomography. Overall, this leads to a more complete view of both the problem and the different approaches.

We begin this section by briefly reviewing the tomography-based approach from the foregoing section. A closer look into the proof of Theorem 2.7 will then directly lead to the study of the ensemble observability problem in a systems theoretic framework. The starting point is the observation that the mapping $s \mapsto \varphi_{x_0}(sv)$ is the characteristic function of the random variable $\langle v, x_0 \rangle$. Real analyticity of $\varphi_{\langle v,x_0 \rangle}$ on $\mathbb{R}$ implies that $\varphi_{\langle v,x_0 \rangle}$ is completely determined by its power series about the origin, which is given by

$$\varphi_{\langle v,x_0 \rangle}(s) = \sum_{p=0}^{\infty} \varphi^{(p)}_{\langle v,x_0 \rangle}(0) \, \frac{s^p}{p!}, \tag{2.15}$$

cf. Section XV.4 in Feller (1971). Furthermore, the derivatives at the origin are

$$\varphi^{(p)}_{\langle v,x_0 \rangle}(0) = i^p \mathbb{E}[\langle v, x_0 \rangle^p],$$

where $\mathbb{E}[\langle v, x_0 \rangle^p]$ is recognized as the pth moment of the distribution of $\langle v, x_0 \rangle$.

In the next subsection, we provide a brief introduction to the notion of moments of probability distributions, as well as the famous moment problem.

Moments of Probability Distributions

Let X be an n-dimensional random vector with a probability distribution $\mathbb{P}$. For a multi-index $\alpha = (\alpha_1, \ldots, \alpha_n)$, i.e. an n-tuple of non-negative integers, we call

$$m_\alpha := \mathbb{E}[X^\alpha] = \mathbb{E}[X_1^{\alpha_1} \cdots X_n^{\alpha_n}] = \int_{\mathbb{R}^n} x_1^{\alpha_1} \cdots x_n^{\alpha_n} \, \mathrm{d}\mathbb{P}(x)$$

a moment of order $|\alpha| = \alpha_1 + \cdots + \alpha_n = p$ of $\mathbb{P}$, if the absolute moment $\int_{\mathbb{R}^n} \|x\|^p \, \mathrm{d}\mathbb{P}(x)$ exists. A classical problem in probability theory is the question as to whether or not a probability distribution is uniquely determined by its moments. This is known as the moment problem (Akhiezer, 1965). If a distribution is determined uniquely by its moments, then the distribution is said to be *moment-determinate*. Moment-determinate distributions, as well as their moments will play a major role in our second approach to the ensemble observability problem.

The moments of $\mathbb{P}_0$ naturally appear in the expansion of the moments $\mathbb{E}[\langle v, x_0\rangle^p]$ using the multinomial theorem, which is given by

$$\mathbb{E}[\langle v, x_0\rangle^p] = \sum_{|\alpha|=p} \binom{p}{\alpha} \mathbb{E}[x_0^\alpha] v^\alpha. \tag{2.16}$$

In particular, from (2.16) we also see that the coefficients of the power series expansion (2.15) are polynomials in the variable v (cf. the claim in the proof of Theorem 2.7). By virtue of this discussion on moments, we conclude that Theorem 2.7 is eventually established by matching the moments of all one-dimensional projections of the considered initial distribution. By the assumption of real analyticity in the previous section, this determines all the one-dimensional projections uniquely, and, by the Cramér-Wold theorem, the considered initial distribution.

Now, we could completely avoid the route over analyticity, by assuming a priori that the considered initial distributions are moment-determinate. Then a second approach would be the consideration of the dynamics of the moments $\mathbb{E}[x^\alpha]$, through which we can aim to characterize the uniqueness of the moment reconstruction problem in terms of observability properties of the moment dynamics. This also seems promising in providing a more systems theoretic treatment for ensemble observability, in contrast to the more analytic treatment in the framework of mathematical tomography.

Before we proceed with studying the dynamics of the moments, it is instructive to note that the class of moment-determinate distributions in fact constitutes a larger class than the class of distributions considered in the previous section.

Proposition 2.12. *The class of moment-determinate distributions contains the class of distributions for which the characteristic functions of all one-dimensional projections are real analytic.*

Proof. We recall that due to real analyticity, the mapping $s \mapsto \varphi_{\langle v,X\rangle}(s)$ can be expanded into a power series of the type (2.15). Now, for two distributions $\mathbb{P}'$ and $\mathbb{P}''$ with the same moments, all their one-dimensional projections for non-zero $v \in \mathbb{R}^n$ have the same moments, cf. (2.16), and are thus equal. By virtue of the Cramér-Wold theorem, we have $\mathbb{P}' = \mathbb{P}''$, i.e. moment-determinacy of the considered distributions. □

Linear Tensor Systems

Our first solution of the ensemble observability problem was essentially based on the idea of matching the coefficients of the power series expansion of the Fourier transforms, which were shown to be related to moments of $\mathbb{P}_0$. Thus, it is natural to directly consider the dynamics of the moments of state and output distributions. Such an approach would, in particular, spare the assumption of real analyticity of the Fourier transform, and moreover seems promising in providing a systems theoretic treatment of the problem.

A useful framework to describe the dynamics of the moments is given by tensor systems (Brockett, 1973), which will be reviewed in the following. In view of (2.16), we define for a vector $x \in \mathbb{R}^n$ the pth tensor power $x^{[p]}$ via the implicit equation

$$\langle v, x\rangle^p = \sum_{|\alpha|=p} \binom{p}{\alpha} v^\alpha x^\alpha =: \langle v^{[p]}, x^{[p]}\rangle. \tag{2.17}$$

More explicitly, $x^{[p]}$ is the vector of weighted powers x^α with $|\alpha| = p$,

$$x^{[p]} = \left(\sqrt{\binom{p}{\alpha}}\, x^\alpha\right)_{|\alpha|=p},$$

where, by convention, the entries of $x^{[p]}$ are ordered lexicographically in a decreasing order according to the multi-indices α. Splitting the multinomial coefficients in (2.17) symmetrically to the vectors v and x is, first of all, a natural definition at this point, and, as we will see, will lead to additional, convenient properties in the tensor system calculus. Moreover, we note that the dimension of $x^{[p]}$ is

$$N(n,p) := \binom{n+p-1}{p}.$$

Of course, the reason for introducing tensor powers $x^{[p]}$ is that their expected values $\mathbb{E}[x^{[p]}]$ are weighted versions of the moments $(m_\alpha)_{|\alpha|=p}$ that we attempt to reconstruct. The tensor power calculus provides a framework which is particularly convenient for our purposes, partly because it was in fact introduced and developed within the study of polynomial systems in control theory, see e.g. Baillieul (1981); Brockett (1973); Dayawansa and Martin (1987); Sastry (1999); Sira-Ramirez (1988).

Given a (static) linear equation $y = Cx$, it can be seen that there is also a linear dependency between the pth powers $y^{[p]}$ and $x^{[p]}$, which we will denote by

$$y^{[p]} = C^{[p]} x^{[p]}.$$

Similarly, if we consider a linear differential equation given by $\frac{d}{dt}x(t) = Ax(t)$, then the evolution of the pth power $x^{[p]}(t)$ is also governed by a linear differential equation which we denote

$$\frac{d}{dt}x^{[p]}(t) = A_{[p]}x^{[p]}(t).$$

Thus, it is noted that there are two types of matrix tensor powers that are of interest, which is reflected in the superscript and subscript in the notation for the tensor power, respectively. Which of the two powers is to be considered depends on whether the underlying equation is static or dynamic.

Example 2.13. Consider a two-dimensional system $\dot{x} = Ax$ with scalar output $y = Cx$. Then we have for the second power of the output

$$y^{[2]} = y^2 = \left(\begin{pmatrix} c_1 & c_2 \end{pmatrix}\begin{pmatrix} x_1 \\ x_2 \end{pmatrix}\right)^2 = \begin{pmatrix} c_1^2 & \sqrt{2}c_1c_2 & c_2^2 \end{pmatrix}\begin{pmatrix} x_1^2 \\ \sqrt{2}x_1x_2 \\ x_2^2 \end{pmatrix}, \tag{2.18}$$

from which we obtain the decomposition $y^{[2]} = (Cx)^{[2]} = C^{[2]}x^{[2]}$ explicitly.

To obtain the decomposition $\frac{d}{dt}x^{[2]} = A_{[2]}x^{[2]}$, we differentiate the entries of $x^{[2]}$, which leads to the linear second order tensor system

$$\frac{d}{dt}x^{[2]} = \begin{pmatrix} 2a_{11} & \sqrt{2}a_{12} & 0 \\ \sqrt{2}a_{21} & a_{11}+a_{22} & \sqrt{2}a_{12} \\ 0 & \sqrt{2}a_{21} & 2a_{22} \end{pmatrix} x^{[2]}, \tag{2.19}$$

and thus also to an explicit description of the matrix $A_{[2]}$.

Tensor powers of matrices satisfy certain properties which we will extensively make use of in the later analysis. First of all, it can be directly verified that for two matrices A and B of compatible dimension, it holds that

$$(AB)^{[p]} = A^{[p]} B^{[p]}.$$

Furthermore, by choosing to split the weights symmetrically in the definition of $x^{[p]}$ in (2.17), for any matrix A, one has

$$(A^\top)^{[p]} = (A^{[p]})^\top,$$

i.e. the operations of transposing and taking the pth tensor power commute.

Having introduced the framework of tensor systems, we are now in the position to derive the following lemma, which establishes a direct connection between the algebraic geometric approach of the previous section and the systems theoretic approach employing tensor systems. We may further refer to this as a *duality* of our two approaches.

Lemma 2.14. *The union* $\bigcup_{t \geq 0} \operatorname{im}(Ce^{At})^\top$ *is contained in the algebraic variety*

$$\{x \in \mathbb{R}^n : \langle a, x^{[p]} \rangle = 0\}$$

if and only if the vector a *is contained in the unobservable subspace of* $(A_{[p]}, C^{[p]})$.

Before we give a proof of Lemma 2.14, we first illustrate its implications on the linear system considered in Example 2.10 of the previous section.

Example 2.15. We reconsider the linear system from Example 2.10 and illustrate the results obtained from the tensor system framework. First of all, we note that for a vector $x \in \mathbb{R}^3$ the second tensor power of x is given by

$$x^{[2]} = \begin{pmatrix} x_1^2 & \sqrt{2}x_1x_2 & \sqrt{2}x_1x_3 & x_2^2 & \sqrt{2}x_2x_3 & x_3^2 \end{pmatrix}^\top.$$

Thus, with the vector

$$a = \begin{pmatrix} 0 & 0 & -\frac{1}{\sqrt{2}} & 1 & 0 & 0 \end{pmatrix}^\top$$

the equation $\langle a, x^{[2]} \rangle = 0$ defines the variety that was introduced in Example 2.10. Furthermore it follows from $Ce^{At} = \begin{pmatrix} 1 & e^{-t} & e^{-2t} \end{pmatrix}$ and the definition of $(Ce^{At})^{[2]}$ via

$$(Ce^{At}x_0)^{[2]} = (Ce^{At})^{[2]} x_0^{[2]}$$

that the second tensor power of Ce^{At} is given by

$$(Ce^{At})^{[2]} = \begin{pmatrix} 1 & \sqrt{2}e^{-t} & \sqrt{2}e^{-2t} & e^{-2t} & \sqrt{2}e^{-3t} & e^{-4t} \end{pmatrix}.$$

Now, a direct computation shows that

$$(Ce^{At})^{[2]} a = 0,$$

for all $t \geq 0$, which is in accordance with Lemma 2.14.

To conclude, the result given in Lemma 2.14 allows us to obtain a *direct* connection between coefficient vectors of the algebraic varieties in Theorem 2.7 and unobservable states of the tensor system: given a vector of the unobservable subspace of a tensor system, this very same vector is also the coefficient vector of a homogeneous polynomial that defines an algebraic variety in which the union (2.10) is contained in.

Proof of Lemma 2.14. The condition that the union $\bigcup_{t\geq 0} \operatorname{im}(Ce^{At})^\top$ is contained in the algebraic variety defined by $\langle a, x^{[p]}\rangle = 0$ is equivalent to

$$\langle a, ((Ce^{At})^\top z)^{[p]}\rangle = 0$$

for all $t \geq 0$ and $z \in \mathbb{R}^m$. Using first $(\tilde{A}\tilde{B})^{[p]} = \tilde{A}^{[p]}\tilde{B}^{[p]}$ and secondly $(\tilde{A}^\top)^{[p]} = (\tilde{A}^{[p]})^\top$, as discussed earlier, we arrive at

$$\langle (Ce^{At})^{[p]}a, z^{[p]}\rangle = 0 \tag{2.20}$$

for all $t \geq 0$ and $z \in \mathbb{R}^m$. Viewing (2.20) as a polynomial equation in z, and taking into account that it holds for all values of its argument $z \in \mathbb{R}^m$, and also all $t \geq 0$, we conclude that (2.20) is equivalent to $a \in \ker(Ce^{At})^{[p]}$ for all $t \geq 0$. Since $(Ce^{At})^{[p]} = C^{[p]}e^{A_{[p]}t}$, the fact that $a \in \ker(Ce^{At})^{[p]}$ for all $t \geq 0$ is equivalent to a being contained in the unobservable subspace of the tensor system $\dot{x}^{[p]}(t) = A_{[p]}x^{[p]}(t)$, $y^{[p]}(t) = C^{[p]}x^{[p]}(t)$. This yields the claim. □

Ensemble Observability Results in a Systems Theoretic Framework

With the introduction of the tensor systems calculus, we are finally in the position to state our main result of this chapter, which gives a unifying, and also more general, sufficient condition for ensemble observability of linear systems based on the observability of the tensor systems.

Theorem 2.16. *The union $\bigcup_{t\geq 0} \operatorname{im}(Ce^{At})^\top$ is contained in no proper algebraic variety if and only if the systems*

$$\begin{aligned} \dot{x}^{[p]}(t) &= A_{[p]}x^{[p]}(t) \\ y^{[p]}(t) &= C^{[p]}x^{[p]}(t) \end{aligned}$$

are observable for all $p \in \mathbb{N}$. Under these equivalent conditions, the system (A, C) is ensemble observable for the class of moment-determinate initial distributions.

To conclude, our main result Theorem 2.16 shows that the tomography approach is in fact in perfect accordance with this moment approach. Through the systems theoretic approach introduced in this section we also learn that moment-determinacy of the considered initial state distributions alone is sufficient, i.e. that the stronger assumption of real analyticity was in fact a technical assumption in Theorem 2.7. Intuitively, this does not come as a surprise, as the idea that given the output distributions we can compute their moments and then reconstruct the moments of the initial state distribution by virtue of observability of the tensor systems is just more direct.

As a corollary of Theorem 2.16, Proposition 2.8 and Corollary 2.9 from the previous section, we have the following result for the special case in which $\operatorname{rank} C = n - 1$.

Corollary 2.17. *If (A, C) is observable and* $\operatorname{rank} C = n-1$, *then (A, C) is ensemble observable for the class of moment-determinate initial distributions. In particular, any two-dimensional observable system (A, C) is ensemble observable for the class of moment-determinate initial distributions.*

In the following example, we further illustrate the concepts of the theoretical framework introduced so far, and more specifically, illuminate the meaning of the algebraic variety defined by (2.10) to the ensemble observability problem. This will, in particular, yield the following result.

Corollary 2.18. *There are systems (A, C) which are observable in the classical sense, but not ensemble observable for the class of moment-determinate initial distributions.*

Example 2.19. We consider again the three-dimensional and decoupled system (2.13) that we discussed within Example 2.10 and Example 2.15, and illustrate the relevance of the non-observability of the second order tensor system to the moments of the considered state distribution. This also demonstrates the consequences of an unobservable second order tensor system for the ensemble observability problem.

First of all, we can switch between weighted and unweighted tensor powers of vectors via the simple change of coordinates

$$x^{[2]} := \begin{pmatrix} 1 & & & & & \\ & \sqrt{2} & & & & \\ & & \sqrt{2} & & & \\ & & & 1 & & \\ & & & & \sqrt{2} & \\ & & & & & 1 \end{pmatrix} \tilde{x}^{[2]}.$$

Furthermore, we recall that if we take expectations in the dynamics of the unweighted monomials $\tilde{x}^{[2]}$, we get exactly the dynamics of the second order moments of the state distribution.

Moreover, by virtue of the above transformation, we can conclude that the transformed coordinate vector

$$\tilde{a} = \begin{pmatrix} 0 & 0 & -\frac{1}{2} & 1 & 0 & 0 \end{pmatrix}^\top$$

is contained in the unobservable subspace of the unweighted tensor system, i.e. the system describing the dynamics of the second order moments. This means that if we add in the covariance matrix

$$\Sigma_{x_0} = \begin{pmatrix} \mathbb{E}[x_1^2] & \mathbb{E}[x_1 x_2] & \mathbb{E}[x_1 x_3] \\ \mathbb{E}[x_1 x_2] & \mathbb{E}[x_2^2] & \mathbb{E}[x_2 x_3] \\ \mathbb{E}[x_1 x_3] & \mathbb{E}[x_2 x_3] & \mathbb{E}[x_3^2] \end{pmatrix} - \mu\mu^\top$$

a sufficiently small $\lambda \in \mathbb{R}$ to the $(2,2)$ element and $-\frac{\lambda}{2}$ to the $(1,3)$ and $(3,1)$ elements respectively such that positive definiteness is preserved for the resulting covariance matrix, then this will not be noticed in the output $\mathbb{E}[y^2(t)]$ of the second order tensor system that describes the evolution of the second order moments.

In the special case of Gaussian distributions, we can thus construct two distinct but indistinguishable initial distributions, disproving ensemble observability of system (2.13). For a concrete example of such indistinguishable initial state distributions, we may consider $\mathbb{P}_0' = \mathcal{N}(\mu, \Sigma')$ and $\mathbb{P}_0'' = \mathcal{N}(\mu, \Sigma'')$, with the positive definite matrices

$$\Sigma' = \begin{pmatrix} \sigma^2 & & \\ & \sigma^2 & \\ & & \sigma^2 \end{pmatrix}, \quad \Sigma'' = \begin{pmatrix} \sigma^2 & 0 & -\frac{\sigma^2}{2} \\ 0 & 2\sigma^2 & 0 \\ -\frac{\sigma^2}{2} & 0 & \sigma^2 \end{pmatrix},$$

i.e. we set $\lambda = \sigma^2$. Given $Ce^{At} = \begin{pmatrix} 1 & e^{-t} & e^{-2t} \end{pmatrix}$, we can compute the variance of the scalar output via the equation

$$\sigma^2_{y(t)} = (Ce^{At})\Sigma_{x_0}(Ce^{At})^\top. \tag{2.21}$$

It is verified via (2.21), that both covariance matrices Σ' and Σ'' lead to the variance

$$\sigma^2_{y(t)} = (1 + e^{-2t} + e^{-4t})\sigma^2.$$

Since the output distribution is also a normal distribution, it is uniquely determined by its mean and variance, which are both the same for $\mathbb{P}'_{y(t)}$ and $\mathbb{P}''_{y(t)}$ by construction. Thus, system (2.13), which is observable in the classical sense, is not ensemble observable, and, in particular, also provides an example for Corollary 2.18.

A Verifiable Condition for Specific Single-Output Systems

In the previous section, we derived a systems theoretic characterization for ensemble observability of linear systems for the class of moment-determinate distributions, which gives a rather complete picture of the ensemble observability problem for linear systems. Although this characterization is to a certain extent comprehensive, in general, verifying the condition that for all $p \in \mathbb{N}$ the pth tensor systems are observable is not feasible, as it requires checking the observability of infinitely many tensor systems. In the following, we derive, for a specific class of systems, a verifiable, necessary and sufficient condition for the observability of all tensor systems. To this end, we will focus on the class of observable single-output systems in which the system matrix has distinct eigenvalues. The resulting condition turns out to be somewhat restrictive, which shows yet again that the class of ensemble observable systems is much smaller than the class of observable systems.

Theorem 2.20. *Consider an observable single-output system (A, C), where the matrix A has distinct eigenvalues $\lambda_1, \ldots, \lambda_n$. Then the tensor systems $(A_{[p]}, C^{[p]})$ are observable for all orders $p \in \mathbb{N}$, if and only if for some $j \in \{1, \ldots, n\}$ the differences*

$$\lambda_i - \lambda_j, \quad i = 1, \ldots, n, \quad i \neq j,$$

are linearly independent over $\mathbb{Q}$.

In the course of proving this result, we will learn that if for some $j \in \{1, \ldots, n\}$ the differences $\lambda_i - \lambda_j, i \neq j$, are linearly independent over $\mathbb{Q}$, then so are the differences for any $j \in \{1, \ldots, n\}$. In other words, if the verification of the linear independence over $\mathbb{Q}$ fails for some choice of $j \in \{1, \ldots, n\}$, then it will also fail for any other choice. Therefore, the verification is independent of the choice of $j \in \{1, \ldots, n\}$.

We note that the condition on the spectrum of A in Theorem 2.20 may be referred to as a *non-resonance condition*, as it is typically considered in the study of integrable Hamiltonian systems, and, more generally, the study of dynamical systems (see e.g. Arnol'd (1983)), though in slightly different forms.

Proof of Theorem 2.20. Recall that any observable single-output system with a system matrix having distinct eigenvalues can be transformed in such a way that the system and output matrices of the transformed system are of the form

$$\tilde{A} = \begin{pmatrix} \lambda_1 & & \\ & \ddots & \\ & & \lambda_n \end{pmatrix},$$
$$\tilde{C} = \begin{pmatrix} \tilde{c}_1 & \ldots & \tilde{c}_n \end{pmatrix},$$

where, necessarily, every entry in $\tilde{C}$ is non-zero. Given the diagonal structure of $\tilde{A}$, it can be seen that the matrix $\tilde{A}_{[p]}$ is then also a diagonal matrix, for which the entries on the diagonal are sums of the form

$$\tilde{\lambda} = \alpha_1\lambda_1 + \cdots + \alpha_n\lambda_n,$$

where α is a multi-index of order p. Moreover, it is important to note that $(\tilde{A}_{[p]}, \tilde{C}^{[p]})$ is indeed also similar to $(A_{[p]}, C^{[p]})$. Now, if some diagonal entries of $\tilde{A}_{[p]}$ have the same value $\tilde{\lambda}$, then in view of a Hautus test for the pth tensor system, computing the difference $\tilde{A}_{[p]} - \tilde{\lambda}I$ will result in a rank loss which is greater than one. The output matrix $\tilde{C}^{[p]}$ being a row vector can, however, only compensate for exactly one rank loss. Therefore, in order to preserve observability in all the higher order (single-output) tensor systems, the possibility of repeated eigenvalues of $\tilde{A}_{[p]}$ needs to be ruled out. The question of whether there exists $p \in \mathbb{N}$ so that

$$\sum_{i=1}^{n} \alpha_i'\lambda_i = \sum_{i=1}^{n} \alpha_i''\lambda_i$$

for different multi-indices α' and α'' of order p is easily seen to be equivalent to the question of whether there exists a vector $z \in \mathbb{Z}^n \backslash \{0\}$ of non-zero integers such that

$$z_1 + \cdots + z_n = 0 \quad \text{and} \quad z_1\lambda_1 + \cdots + z_n\lambda_n = 0. \tag{2.22}$$

It remains only to show that this is, in turn, equivalent to the claimed linear independence condition on the spectrum of A. To this end, suppose there exists $z \in \mathbb{Z}^n \backslash \{0\}$ such that (2.22) holds. Given an arbitrary index $j \in \{1, \ldots, n\}$, we have

$$\sum_{i \neq j} z_i(\lambda_i - \lambda_j) = \sum_{i \neq j} z_i\lambda_i - \Big(\sum_{i \neq j} z_i\Big)\lambda_j = \sum_{i=1}^{n} z_i\lambda_i = 0. \tag{2.23}$$

Therein, we solved the equation $z_1 + \cdots + z_n = 0$ for $z_j = -\sum_{i \neq j} z_i$ in the second equality. Thus, there exist integer, and hence also rational, coefficients q_i such that

$$\sum_{i \neq j} q_i(\lambda_i - \lambda_j) = 0. \tag{2.24}$$

Conversely, suppose that (2.24) holds for some non-zero rational coefficients q_i. There a rescaling (by multiplying with all the denominators) leads to the consideration of the foregoing case of integer coefficients, cf. (2.23), from which a vector $z \in \mathbb{Z}^n \backslash \{0\}$ satisfying (2.22) is readily constructed. This shows that, for all $j \in \{1, \ldots, n\}$, the equivalence of the existence of $z \in \mathbb{Z}^n \backslash \{0\}$ such that (2.22) holds and the existence of non-zero rational coefficients q_i such that (2.24) holds. □

Theorem 2.20 has the following slightly simpler reading though less general corollary.

Corollary 2.21. *Consider an observable single-output system (A, C), where A has distinct eigenvalues $\lambda_1, \ldots, \lambda_n$. Then the tensor systems $(A_{[p]}, C^{[p]})$ are observable for all orders $p \in \mathbb{N}$, if all non-zero eigenvalues $\lambda_i \neq 0$ are* linearly independent over $\mathbb{Q}$.

Proof. The crucial part is that for the considered system class, there can be at most one eigenvalue that is zero. If one eigenvalue is zero, we may choose $\lambda_j = 0$ and apply Theorem 2.20. If the system does not have an eigenvalue at zero, linear independence of λ_i for all $i \in \{1, \ldots, n\}$ trivially implies the linear independence of the differences $\lambda_i - \lambda_j$. □

To demonstrate the applicability of the theoretical framework developed so far within this chapter, we present in the following a treatment of the ensemble observability problem under the additional assumption that the individual components of the initial state distribution are independent.

Incorporating Independence of Initial State Components

In this subsection, we examine what is to be gained if one has knowledge about the components of the random vector x_0 being independent. As before, our standing assumption is moment-determinacy of the considered probability distributions. We recall that for these distributions, the ensemble observability problem is equivalent to the reconstruction of all the moments of the distribution.

We assume that the components of the random initial state x_0 are independent, i.e. that the density of the initial state can be factored as

$$p_0(x) = p_{0,1}(x_1) \cdots p_{0,n}(x_n)$$

or, in other words, that the components $x_{0,i}$ of the random vector x_0 are independent. The consideration of this assumption is relevant for practical problems which frequently admit such an independence between the state components. Moreover, the analysis provided in the following serves as an illustration of the theoretical framework that we developed up to this point.

To simplify matters in our consideration, the first step is to consider *cumulants* rather than moments. It should be kept in mind that if moments exist, then cumulants also exist and that one can then directly compute moments from cumulants. Recall that for a random variable Y, the cumulants κ_p are defined by a power series expansion of the so-called cumulant-generating function

$$g(t) = \log \mathbb{E}[e^{tY}] = \sum_{p=1}^{\infty} \kappa_p \frac{t^p}{p!}.$$

We further recall that the pth cumulant is homogeneous of degree p in the sense that

$$\kappa_p(cY) = c^p \kappa_p(Y)$$

for all $c \in \mathbb{R}$. Furthermore, if Y' and Y'' are two independent random variables, one has the additivity of cumulants

$$\kappa_p(Y' + Y'') = \kappa_p(Y') + \kappa_p(Y'').$$

The fact that independence can be naturally incorporated via additivity is what makes the consideration of cumulants rather than moments particularly attractive.

Now, using the independence of initial state components $x_{0,i}$, where $i \in \{1, \ldots, n\}$, and homogeneity of the cumulants, we have

$$\kappa_p(\langle s, Ce^{At}x_0 \rangle) = \sum_{i=1}^{n} ((Ce^{At})^\top s)_i^p \kappa_p(x_{0,i}) \tag{2.25}$$

for arbitrary non-zero $s \in \mathbb{R}^m$. Therein, the left-hand side is known, and $\kappa_p(x_{0,i})$ are the unknowns that we would like to solve for. We further note that (2.25) can be uniquely solved for the pth cumulants of x_0, if the union $\bigcup_{t \geq 0} \operatorname{im}(Ce^{At})^\top$ is not contained in a proper algebraic variety of the form

$$\tilde{a}_1 x_1^p + \cdots + \tilde{a}_n x_n^p = 0. \tag{2.26}$$

Thus, we conclude that by considering an independence assumption for the considered initial state distributions, we ultimately shrink the class of algebraic varieties to be considered in the richness condition to those algebraic varieties defined by polynomials of degree p, which do not have cross terms. Here we conveniently say that a polynomial of degree p does not have a cross-term, if all monomials occurring in the polynomial are of the form $x^\alpha = x_i^{|\alpha|}$. From the tensor system viewpoint, the union $\bigcup_{t \geq 0} \operatorname{im}(Ce^{At})^\top$ is not contained in an algebraic variety of the form (2.26) if and only if the *intersection* of the unobservable subspace of $(A_{[p]}, C^{[p]})$ *with* the subspace

$$\{a \in \mathbb{R}^{N(n,p)} : a_i = 0 \Leftrightarrow i\text{th entry of } x^{[p]} \text{ is a cross-term}\} \tag{2.27}$$

is trivial. This means that in view of testing observability of the tensor systems via a Hautus test, we do not need to consider every eigenvector of $A_{[p]}$, but only those that additionally lie in the set (2.27).

Example 2.22. Example 2.19 shows that system (2.13) is a system which becomes ensemble observable with the additional assumption of independence. This can be seen from inspecting the unobservable subspace which is spanned by

$$a = \begin{pmatrix} 0 & 0 & -\frac{1}{\sqrt{2}} & 1 & 0 & 0 \end{pmatrix}^\top .$$

Thus the unobservable subspace is not contained in a set of the form (2.27), which, in the three-dimensional case, is the linear span of the set containing the standard basis vectors $e_1, e_4, e_6 \in \mathbb{R}^6$. In particular, the intersection of both subspaces is trivial.

Another way to see that system (2.13) becomes ensemble observable under the independence assumption is through the fact that $\lambda = 0$ must necessarily hold in Example 2.19 in order to preserve the independence that is reflected in the diagonal structure of the covariance matrix. This obstructs us from constructing another indistinguishable initial distribution which satisfies the independence assumption.

Although it is clear from the foregoing discussion and the above example that it is in principle easier to reconstruct an initial state distribution which is known to satisfy an independence assumption, it is in general still not the case that observability of (A, C) alone implies ensemble observability. In the following example, we show this by explicitly constructing a counterexample.

Example 2.23. Given a three-dimensional single-output system (A, C), it can be verified that the dynamics of the second order moments

$$(m_\alpha)_{|\alpha|=2} := \begin{pmatrix} \mathbb{E}[x_1^2] & \mathbb{E}[x_1 x_2] & \mathbb{E}[x_1 x_3] & \mathbb{E}[x_2^2] & \mathbb{E}[x_2 x_3] & \mathbb{E}[x_3^2] \end{pmatrix}^\top ,$$

are described by the system matrix

$$\tilde{A}_{[2]} := \begin{pmatrix} 2a_{11} & 2a_{12} & 2a_{13} & 0 & 0 & 0 \\ a_{21} & a_{11}+a_{22} & a_{23} & a_{12} & a_{13} & 0 \\ a_{31} & a_{32} & a_{11}+a_{33} & 0 & a_{12} & a_{13} \\ 0 & 2a_{21} & 0 & 2a_{22} & 2a_{23} & 0 \\ 0 & a_{31} & a_{21} & a_{32} & a_{22}+a_{33} & a_{23} \\ 0 & 0 & 2a_{31} & 0 & 2a_{32} & 2a_{33} \end{pmatrix} .$$

Here we dropped the weights in the definition of the second power of x for simplicity, which will not alter the result of this qualitative observability analysis. The second order moment of the output, $\mathbb{E}[y^2]$, is furthermore related to the second order moments of the state by the output matrix

$$\tilde{C}^{[2]} := \begin{pmatrix} c_1^2 & 2c_1 c_2 & 2c_1 c_3 & c_2^2 & 2c_2 c_3 & c_3^2 \end{pmatrix} .$$

Now, for the ensemble observability analysis under the independence assumption, we only need to consider eigenvectors where the second, third and fifth entries are zero. To construct a counterexample, we need to find an observable (A, C) failing such a constrained Hautus test for the second order tensor system.

In order to fail such a constrained Hautus test, we need to be able to find a solution $\tilde{v}$ to the eigenvalue problem

$$\tilde{A}_{[2]}\tilde{v} = \begin{pmatrix} 2a_{11}v_1 \\ a_{21}v_1 + a_{12}v_2 \\ a_{31}v_1 + a_{13}v_3 \\ 2a_{22}v_2 \\ a_{32}v_2 + a_{23}v_3 \\ 2a_{33}v_3 \end{pmatrix} = \lambda \begin{pmatrix} v_1 \\ 0 \\ 0 \\ v_2 \\ 0 \\ v_3 \end{pmatrix} = \lambda\tilde{v}$$

subject to the constraint $\tilde{C}^{[2]}\tilde{v} = 0$. First of all, it is seen from the eigenvalue problem that $2a_{ii} = \lambda$ needs to hold. Now, if we choose $v_1 = 1, v_2 = 1$ and $v_3 = -1$, this results in the equations

$$a_{21} + a_{12} = 0, \quad a_{31} - a_{13} = 0, \quad a_{32} - a_{23} = 0.$$

Moreover, if we choose $C = \begin{pmatrix} 1 & 1 & \sqrt{2} \end{pmatrix}$, then $\tilde{C}^{[2]}\tilde{v} = 0$. Based on these considerations, we consider the system

$$\dot{x}(t) = \left(\begin{array}{cc|c} 0 & 1 & 0 \\ -1 & 0 & 0 \\ \hline 0 & 0 & 0 \end{array}\right) x(t), \quad y(t) = \left(\begin{array}{cc|c} 1 & 1 & \sqrt{2} \end{array}\right) x(t),$$

which can be verified to be observable, but is in fact constructed such that the resulting second order tensor system $(\tilde{A}_{[2]}, \tilde{C}^{[2]})$ fails the constrained Hautus test.

For two concrete indistinguishable initial distributions, we consider two normal distributions $\mathbb{P}_0' = \mathcal{N}(\mu, \Sigma')$ and $\mathbb{P}_0'' = \mathcal{N}(\mu, \Sigma'')$ with the same mean and covariance matrices

$$\Sigma' = \begin{pmatrix} \sigma^2 & & \\ & \sigma^2 & \\ & & \sigma^2 \end{pmatrix}, \quad \Sigma'' = \begin{pmatrix} \frac{3}{2}\sigma^2 & & \\ & \frac{3}{2}\sigma^2 & \\ & & \frac{1}{2}\sigma^2 \end{pmatrix},$$

which we constructed via $\tilde{v}$ in the unobservable subspace of $(\tilde{A}_{[2]}, \tilde{C}^{[2]})$, cf. Example 2.19. Using again the relation

$$\sigma_{y(t)}^2 = (Ce^{At})\Sigma_{x_0}(Ce^{At})^\top$$

and the fact that

$$Ce^{At} = \begin{pmatrix} \cos(t) - \sin(t) & \cos(t) + \sin(t) & \sqrt{2} \end{pmatrix},$$

we obtain an output variance of $\sigma_{y(t)}^2 = 4\sigma^2$ for both initial distributions.

2.5 Ensemble Observability of Nonlinear Systems

For a first treatment of the ensemble observability problem, we restricted our attention entirely to the class of linear systems. In this case, we were able to reveal an intimate connection to the theory of mathematical tomography, and eventually obtained both algebraic geometric and systems theoretic characterizations of ensemble observability. It is also of interest to extend the scope of our framework to nonlinear systems, which is the subject of this section. To this end, we adopt the nonlinear model

$$\begin{aligned} \dot{x}(t) &= f(x(t)), \quad x(0) = x_0, \\ y(t) &= h(x(t)), \end{aligned}$$

with sufficiently smooth functions f and h, and a random vector $x_0 \sim \mathbb{P}_0$ for a given initial state distribution $\mathbb{P}_0$. Analogously to the linear case, in the ensemble observability problem for nonlinear systems we ask under which conditions we can reconstruct the initial state distribution $\mathbb{P}_0$ when given the evolution of the distribution of outputs $\mathbb{P}_{y(t)}$. Furthermore, we are interested in practical reconstruction techniques for this problem.

Even though we expect the nonlinear case to be quite different from the linear one, the basic pushforward equation carries over analogously. In the nonlinear case, the output distribution is still given in terms of the pushforward of the initial state distribution, i.e. it holds that

$$\mathbb{P}_{y(t)}(B_y) = \mathbb{P}_0((h \circ \Phi_t)^{-1}(B_y)) = \int_{(h \circ \Phi_t)^{-1}(B_y)} p_0(x)\, \mathrm{d}x, \tag{2.28}$$

where Φ_t denotes the flow of the nonlinear system.

Again, in the ensemble observability problem, we are given the distributions $\mathbb{P}_{y(t)}$ and would like to infer the initial state distribution $\mathbb{P}_0$. By virtue of the pushforward relation (2.28), the ensemble observability problem is thus to infer an unknown density p_0 from its integral values along sets of the form $(h \circ \Phi_t)^{-1}(B_y)$, where $t \geq 0$ and $B_y \subset \mathbb{R}^m$ are allowed to vary. This problem description is illustrated in Figure 2.7.

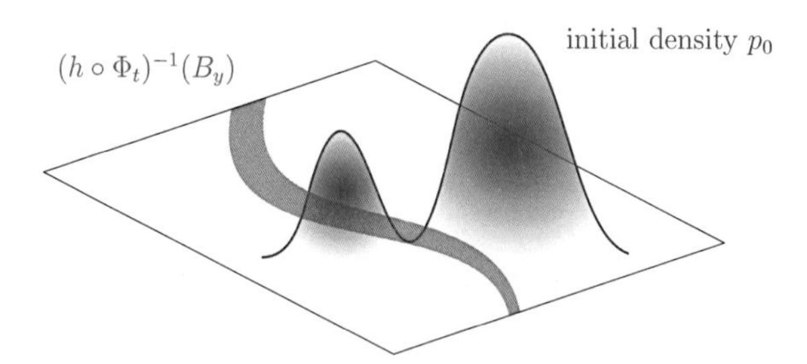

Figure 2.7: This figure illustrates the problem at the core of the ensemble observability problem for nonlinear systems, which is the reconstruction of an unknown density p_0 from its integral values along the curved strips $(h \circ \Phi_t)^{-1}(B_y)$.

The situation is quite different when attempting to formulate the pushforward relation in terms of surface integrals by the idea of taking the width of a strip to zero. In the linear case, we were able to reformulate the ensemble observability problem as a standard tomography problem in which the given output densities $p_{y(t)}$ were linked to surface integrals of the unknown initial density p_0 over affine subspaces defined by $Ce^{At}x = y$. This reformulation was based on the coarea formula, i.e. the relation

$$\int_{H^{-1}(B_y)} p(x)|J_m H(x)| \, \mathrm{d}x = \int_{B_y} \left(\int_{H^{-1}(\{y\})} p(x) \, \mathrm{d}S \right) \mathrm{d}y$$

for a general possibly nonlinear mapping $H : \mathbb{R}^n \to \mathbb{R}^m$. In the linear ensemble observability problem, we had $H(x) = Ce^{At}x$, of which the m-dimensional Jacobian is independent of the state. This allowed us to shift the Jacobian term from the integral on the left-hand side to the right-hand side and ultimately, to obtain a closed formula for the output densities as normalized integrals of the initial state density over affine subspaces defined by $Ce^{At}x = y$. The fact that $|J_m H(x)|$ with $H(x) = (h \circ \Phi_t)(x)$ is always independent of the state, however, holds for linear systems only. That is, in the nonlinear case, the Jacobian $|J_m H|$ will in general be state-dependent, so that, in particular, an explicit relation in terms of surface integrals over manifolds defined by the nonlinear equation $(h \circ \Phi_t)(x) = y$ is *not valid*. In Figure 2.7, this state dependency of the Jacobian in the general nonlinear case is hinted in terms of a non-uniform width of the curved strip, which also gives a clue as to why the process of cutting down the width of such a strip to arrive at a manifold given by $(h \circ \Phi_t)(x) = y$ is more subtle.

For an illustration of the nonlinear ensemble observability problem via the pushforward equation, i.e. the tomography perspective, we consider the following example.

Example 2.24. For a concrete example, we consider an ensemble of nonlinear systems given by

$$\begin{aligned} \dot{x}_1 &= x_2, \\ \dot{x}_2 &= -4x_1 + x_1^2, \end{aligned} \tag{2.29}$$

as considered in Brockett (2010) within the study of a broadcast stabilization problem in a slightly more general form. This nonlinear system admits periodic orbits close to the origin, which rotate clockwise. Moreover, we shall consider the output $y = x_1$. Again, in the ensemble observability problem, we observe the output distribution $\mathbb{P}_{y(t)}$ over time, which, in this example, results from marginalizing the propagation of the initial state distribution $\mathbb{P}_0$ over the second coordinate.

The top row of Figure 2.8 illustrates the phase portrait of the nonlinear oscillator (2.29), as well as the propagation of a certain bimodal initial state distribution with the flow for two given time points. The measured output distributions corresponding to the two time points are illustrated in the second row of Figure 2.8. The third row of Figure 2.8 illustrates the initial distribution that underlies both propagated state distributions in the first row. Furthermore, the red lines indicate the preimages of the form $(h \circ \Phi_t)^{-1}(B_y)$. By virtue of (2.28), the measured output distributions provide the values of the integrals of p_0 over the curved strips in the third row of Figure 2.8.

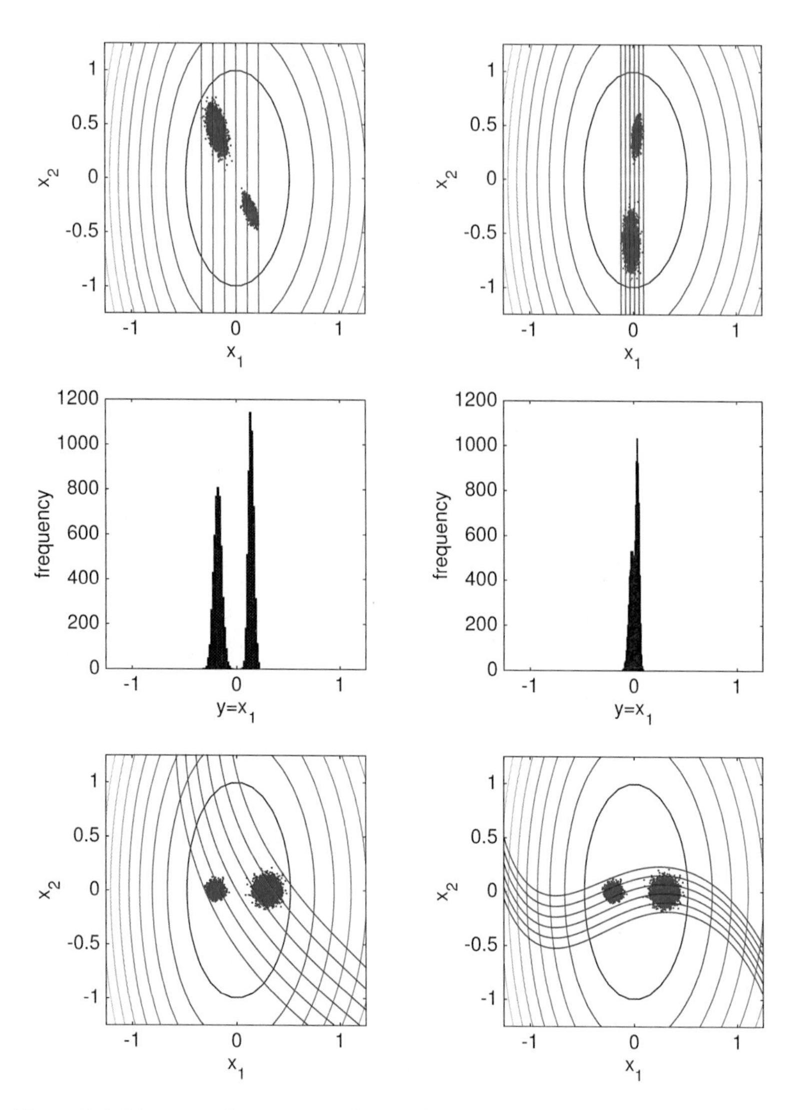

Figure 2.8: Top row: the propagated state distributions for a given initial state distribution at two different time points, as well as the level sets of the output $y = h(x) = x_1$. Middle row: measured output distributions in the form of histograms corresponding to the plots in the first row. Bottom row: the initial state distribution with the transported level sets of the output $y = x_1$ via the flow of the nonlinear system backwards in time.

In this geometric illustration it is interesting to note that the two plots in the last row of Figure 2.8 can in fact be obtained by applying the reverse flow to their corresponding plots in the first row, i.e. by applying the reverse flow on both the propagated state distribution and the sets $h^{-1}(B_y)$ indicated by the red lines. Indeed, this property is in accordance with an intermediate step of (2.28) given by

$$\int_{h^{-1}(B_y)} p_t(x)\,\mathrm{d}x = \int_{\Phi_{-t}(h^{-1}(B_y))} p_0(x)\,\mathrm{d}x, \tag{2.30}$$

where p_t denotes the density of the state distribution at time t. Intuitively speaking, (2.30) reflects the fact that the "mass" obtained from integrating the density p_t over the set $h^{-1}(B_y)$ is the same as the "mass" obtained from integrating the (backwards) propagation of p_t, i.e. p_0, over the (backwards) propagation of the set $h^{-1}(B_y)$. This situation is related to the so-called measure preserving property inherent to the considered setup. One should also note that for this reason, (2.30) is sometimes referred to as a continuity equation, and that the aforementioned continuity property serves as the basic definition of the operator semigroup $(P_t)_{t\geq 0}$ that describes the evolution of a density with the flow of a dynamical system, cf. Lasota and Mackey (1994). The infinitesimal generator of this semigroup is the partial differential equation

$$\frac{\partial}{\partial t}p(t,x) = -\operatorname{div}(p(t,x)f(x)), \quad p(0,x) = p_0(x),$$

which is the Liouville equation that we briefly mentioned in the introduction.

We proceed with a discussion of the two approaches to the ensemble observability in the linear case, given by the Fourier-based and moment-based approaches, respectively, as potential techniques for the analysis of the nonlinear case. While the Fourier-based technique of the tomography viewpoint does not seem to easily carry over to the nonlinear case, the moment-based methodologies can still be readily employed, at least in the case of polynomial nonlinearities. Unlike in the linear case, however, the dynamics of the pth moments does not only depend on the pth moments, but on higher order moments as well. This situation is typically referred to as a non-closure of the moment equations.

Example 2.25. We reconsider the nonlinear oscillator (2.29) and illustrate the moment-based approach in the nonlinear case. First of all, the differential equation for the evolution of the first order moments is

$$\frac{d}{dt}\begin{pmatrix}\mathbb{E}[x_1]\\ \mathbb{E}[x_2]\end{pmatrix} = \begin{pmatrix}0 & 1\\ -4 & 0\end{pmatrix}\begin{pmatrix}\mathbb{E}[x_1]\\ \mathbb{E}[x_2]\end{pmatrix} + \begin{pmatrix}0\\ \mathbb{E}[x_1^2]\end{pmatrix}$$

and for the second order moments we have

$$\frac{d}{dt}\begin{pmatrix}\mathbb{E}[x_1^2]\\ \mathbb{E}[x_1x_2]\\ \mathbb{E}[x_2^2]\end{pmatrix} = \begin{pmatrix}0 & 2 & 0\\ -4 & 0 & 1\\ 0 & -1 & 0\end{pmatrix}\begin{pmatrix}\mathbb{E}[x_1^2]\\ \mathbb{E}[x_1x_2]\\ \mathbb{E}[x_2^2]\end{pmatrix} + \begin{pmatrix}0\\ \mathbb{E}[x_1^3]\\ 2\mathbb{E}[x_1^2x_2]\end{pmatrix},$$

which showcases the non-closure for this example.

One attempt to actually prove ensemble observability for the considered nonlinear oscillator (2.29) is to consider the equality $\mathbb{E}[(h \circ \Phi_t)(x_0')^p] \equiv \mathbb{E}[(h \circ \Phi_t)(x_0'')^p]$ for all $p \in \mathbb{N}$ (and all their derivatives) and to conclude from this that the moments of the corresponding random vectors x_0' and x_0'' are equal for all orders. This is, as will be discussed in the following, equivalent to the consideration of the vanishing of the signals $t \mapsto \mathbb{E}[y^p(t)] = \mathbb{E}[x_1^p(t)]$ and all its derivatives, and to attempt to conclude from this that the signal $t \mapsto \mathbb{E}[x^{[p]}(t)]$ vanishes identically. Of course, the description of this ansatz involves a slight abuse of notation in that we refer to $\mathbb{E}[x^{[p]}]$ here not as the moments of an actual distribution but rather as the states of the *linear* infinite-dimensional system

$$\frac{d}{dt}\begin{pmatrix} \mathbb{E}[x_1] \\ \mathbb{E}[x_2] \\ \hline \mathbb{E}[x_1^2] \\ \mathbb{E}[x_1 x_2] \\ \mathbb{E}[x_2^2] \\ \hline \vdots \end{pmatrix} = \left(\begin{array}{cc|ccc|cccc|c} 0 & 1 & & & & & & & & \\ -4 & 0 & 1 & & & & & & & \\ \hline & & 0 & 2 & 0 & & & & & \\ & & -4 & 0 & 1 & 1 & & & & \\ & & 0 & -1 & 0 & 0 & 2 & & & \\ \hline & & & & & \ddots & & & & \ddots \end{array}\right) \begin{pmatrix} \mathbb{E}[x_1] \\ \mathbb{E}[x_2] \\ \hline \mathbb{E}[x_1^2] \\ \mathbb{E}[x_1 x_2] \\ \mathbb{E}[x_2^2] \\ \hline \mathbb{E}[x_1^3] \\ \mathbb{E}[x_1^2 x_2] \\ \mathbb{E}[x_1 x_2^2] \\ \mathbb{E}[x_2^3] \\ \hline \vdots \end{pmatrix}$$

that describes the evolution of the moments. This system is, in particular, defined on the space of sequences. Moreover, we note that the general idea employed here is essentially in the same spirit as the so-called Carleman embedding technique for nonlinear systems (Carleman, 1932), where one "embeds" a finite-dimensional real analytic nonlinear system as an infinite-dimensional *linear* system defined on the space of real sequences, see e.g. Brockett (2014) and references therein. In the linear ensemble observability problem, the occurring infinite-dimensionality was not an issue, as the completely decoupled structure of the moment dynamics allowed us to consider the tensor systems $(A_{[p]}, C^{[p]})$ separately.

In the following, we outline a proof for the ensemble observability of the considered example by moment-based methods. First of all, by linearity of the moment dynamics, we can incorporate the fact that $y = x_1$ by the vanishing $\mathbb{E}[x_1^p] \equiv 0$ for all $p \in \mathbb{N}$. Moreover, for the first order moments, we derive $\frac{d}{dt}\mathbb{E}[x_1] = \mathbb{E}[x_2] \equiv 0$. Similarly, by considering $\mathbb{E}[x_1^2] \equiv 0$ and differentiating

$$\frac{d}{dt}\mathbb{E}[x_1^2] = 2\mathbb{E}[x_1 x_2],$$

we conclude that $\mathbb{E}[x_1 x_2] \equiv 0$. Furthermore, the vanishing $\mathbb{E}[x_2^2] \equiv 0$ follows from

$$\frac{d}{dt}\mathbb{E}[x_1 x_2] = \mathbb{E}[x_2^2] - 4\mathbb{E}[x_1^2] + \mathbb{E}[x_1^3] \equiv 0.$$

We next generalize the foregoing consideration for the first and second order moments to arbitrary moments $\mathbb{E}[x_1^{\alpha_1} x_2^{\alpha_2}] \equiv 0$ with $\alpha \in \mathbb{N}^2$. This eventually establishes ensemble observability of the nonlinear oscillator for the class of moment-determinate distributions. Analogously to the cases of first and second order moments, we start with considering

$$\frac{d}{dt}\mathbb{E}[x_1^p] = p\mathbb{E}[x_1^{p-1} x_2] \equiv 0$$

and then repeatedly use the rule

$$\frac{d}{dt}\mathbb{E}\big[x_1^{p-k} x_2^k\big] = (p-k)\mathbb{E}\big[x_1^{p-(k+1)} x_2^{k+1}\big] - 4k\mathbb{E}\big[x_1^{p-(k-1)} x_2^{k-1}\big] + k\mathbb{E}\big[x_1^{(p+1)-(k-1)} x_2^{k-1}\big]$$

for all $k \in \{1, \dots, p-1\}$. In each step of the repeated process of differentiating $\mathbb{E}\big[x_1^{p-k} x_2^k\big]$, the second term is associated with the predecessor of order $k-1$ in the chain of differentiations and therefore can be taken to be zero. The third term is associated with the chain of differentiations for the moments of order $p+1$, and, by going through the chain of differentiation that is initialized with $\mathbb{E}[x_1^{p+1}] \equiv 0$, can be assumed to be zero as well. Therefore, we conclude that $\mathbb{E}\big[x_1^{p-(k+1)} x_2^{k+1}\big] \equiv 0$, which is also the lexicographical successor of $\mathbb{E}\big[x_1^{p-k} x_2^k\big]$ in the vector of pth moments.

It is worthwhile to note that the result considered in the foregoing example can be generalized to the following two-dimensional (Hamiltonian) systems of the form

$$\begin{aligned} \dot{x}_1 &= x_2, \\ \dot{x}_2 &= q(x_1), \end{aligned}$$

where q is an arbitrary polynomial, and the choice of the output $y = x_1$. In fact, even more generally, one may relax the assumption of q being a polynomial by q being a real analytic function. Thus the above system may be viewed as a normal form for ensemble observability in the two-dimensional case. As already witnessed in the linear case, however, it must be acknowledged that the two-dimensional case appears to be rather special. Even in the linear ensemble observability problem, the situation for higher-dimensional systems became more subtle. The following example illustrates the more delicate situation for an ensemble of heterogeneous harmonic oscillators which can be formulated in terms of the three-dimensional nonlinear system

$$\begin{aligned} \dot{x}_1 &= x_2 x_3, \\ \dot{x}_2 &= -x_1 x_3, \\ \dot{x}_3 &= 0. \end{aligned}$$

This polynomial system models a harmonic oscillator $\dot{x}_1 = \theta x_2, \dot{x}_2 = -\theta x_1$, where we introduced a third state variable $x_3 = \theta$ as the constant, but heterogeneous frequency. The assumption $x_0 \sim \mathbb{P}_0$ thus models a distribution in both phase and frequency of the family of two-dimensional oscillators. Suppose further that the output measurements consist of $y_1 = x_1$ and $y_2 = x_2$. A direct calculation, similar to that of the foregoing example, reveals that for certain moments $\mathbb{E}[x^\alpha]$ no statements about the uniqueness of the reconstruction problem can be made.

In fact, by considering a simple example which involves a uniform distribution supported on a torus that is radially symmetric about the x_3-axis, we see that shifting the distribution along the x_3-axis yields the same invariant uniform distribution for the output distribution. This shows that in the course of studying ensemble observability of higher dimensional nonlinear systems, one has to further restrict the class of considered initial distributions in order to obtain practically relevant results.

Practical Reconstruction Based on Tomography Methods

For the practical reconstruction of the initial state distribution, we note that we can once again leverage the geometric viewpoint. The general idea for the practical reconstruction, as presented for linear systems at the end of Section 2.3, is still feasible in the nonlinear case and can in fact be treated analogously, i.e. by employing the basic idea of Algebraic Reconstruction Techniques. The only additional difficulty in the nonlinear case lies in the practical computation of the weights that incorporate the area of a curved strip $(h \circ \Phi_t)^{-1}(B_y)$ within the different pixels S_i, which are needed in view of the approximation

$$\mathbb{P}_0((h \circ \Phi_t)^{-1}(B_y)) \approx \sum_{i=1}^{N} p_i \frac{|S_i \cap (h \circ \Phi_t)^{-1}(B_y)|}{|S_i|}.$$

These weights can, however, be practically computed by exploiting the fact that

$$(h \circ \Phi_t)^{-1}(B_y) = \Phi_{-t}(h^{-1}(B_y)),$$

i.e. that the set $(h \circ \Phi_t)^{-1}(B_y)$ is the result of propagating the preimage $h^{-1}(B_y)$ with the flow of the nonlinear system backwards in time. The idea is then to discretize the boundary of the preimages $h^{-1}(B_y)$ with grid points in $\mathbb{R}^n$ and to transport these grid points with the flow backwards in time. This provides an approximation of the set $(h \circ \Phi_t)^{-1}(B_y)$ in terms of its boundary, from which the portion that $(h \circ \Phi_t)^{-1}(B_y)$ occupies in a given pixel S_i can be computed. We note that, by this discussion, the method described here is restricted to output mappings which allow for an efficient discretization of its level surfaces. We refer to Zeng and Allgöwer (2015) for more details on the actual implementation and a practical reconstruction for the example of the nonlinear oscillator, the presentation of which is beyond the scope of this thesis.

2.6 Summary and Discussion

In this chapter, we introduced the systems theoretic concept of ensemble observability of dynamical systems and provided a theoretical framework in which this concept was studied and characterized. We first illustrated the ensemble observability problem for linear systems from different viewpoints, of which one revealed a natural connection to mathematical tomography. This fundamental connection eventually brought together observability and mathematical tomography, two ideas that had, interestingly, emerged at approximately the same time in the early 1960s, though with different scopes.

We first pursued the study of the ensemble observability problem using the framework provided by mathematical tomography. This eventually led to a characterization of ensemble observability, for a specific class of initial distributions, in terms of a geometric richness condition given in Theorem 2.7. More precisely, the "directions" generated by the linear system need to be sufficiently rich in the sense that they are not contained in a proper projective variety, or in other words need to form an open set in the Zariski topology. The framework of tomography was not only useful for theoretical studies, but was also leveraged as a method for practically reconstructing initial state distributions. We discussed the direct application of so-called Algebraic Reconstruction Techniques from tomography and illustrated these on an exemplary linear system. One of the advantages of viewing the ensemble observability problem as a tomography problem is that this viewpoint captures a more "global" description, from which one would, in particular, expect a numerically more "stable" reconstruction.

In the second part of this chapter, we pursued another, more systems theoretic approach for characterizing ensemble observability of linear systems. Therein, the basic idea was to consider the dynamical system that describes the evolution of the moments of the state and output distributions. Under the assumption that the considered distributions are moment-determinate, a given linear system is ensemble observable if all the dynamical systems describing the evolution of the moments are observable. Moreover, we presented a comprehensive discussion on the link between this second approach and the tomography approach. A key tool in the systems theoretic approach were tensor systems $(A_{[p]}, C^{[p]})$, which provide a convenient framework for studying the dynamics of the moments of state and output distributions of order p. The main result of the second part of this chapter is Theorem 2.16, which gives a characterization of ensemble observability of linear systems for the class of moment-determinate initial distributions in terms of the observability of an infinite family of associated linear tensor systems. Furthermore, Theorem 2.16 provides a direct connection to the richness condition in Theorem 2.7 of the first part of this chapter. We used the framework of tensor systems to establish some further, more specialized results concerning ensemble observability of linear systems. For instance, Theorem 2.20 shows that for the class of observable single-output systems with distinct eigenvalues, the observability of all tensor systems $(A_{[p]}, C^{[p]})$ is equivalent to a specific linear independence condition on the eigenvalues of A, which may be described as a non-resonance condition on the spectrum of A.

In the last part of this chapter, we considered the ensemble observability problem for nonlinear systems. We showed that in this case, a tomography-based viewpoint is still valid, though it actually leads to a nonstandard nonlinear tomography problem. This perspective was illustrated by means of the example of a nonlinear oscillator. We showed that the practical state reconstruction techniques employing this viewpoint are applicable to nonlinear and linear systems alike. However, from a theoretical point of view, the nonlinear case is not yet fully understood. Particular difficulties are the breakdown of the Fourier-based techniques which we could employ in the linear case and the problem of non-closure of the moment equations in the moment-based framework, respectively. For the example of the specific nonlinear oscillator, however, we were eventually able to prove ensemble observability using moment-based techniques in a manner reminiscent of the Carleman embedding technique for classical nonlinear systems.

3 Sampled Observability of Discrete Linear Ensembles

In the previous chapter, we considered the question under which conditions a continuous distribution of initial states can be determined from observing the complete time evolution of a corresponding distribution of outputs. Our analysis eventually led to both algebraic geometric and systems theoretic characterizations of ensemble observability of linear systems. The former characterization was through a geometric richness condition formulated mathematically in terms of non-membership of the available measurement directions in projective varieties, while the latter characterization was through the observability of higher order tensor systems. Although these characterizations are definite, it is not always easy to understand them intuitively.

It is hence only natural to consider the, perhaps more tangible, discrete counterpart which results from replacing the absolutely continuous distribution with a discrete distribution comprising a finite number of support points and the continuous measurement times with discrete ones. The consideration of this discrete version gives rise to further geometric and algebraic structures. We study these from different viewpoints, which in particular highlight the connection to the continuous ensemble observability of linear systems. On the other hand, the sampled observability problem for discrete ensembles may also be viewed independently of the continuous ensemble observability problem, as it poses a basic sampled-data control theoretic question in itself. From this point of view, one considers a set of N linear systems of which outputs are measured at discrete time points. The premise of ensemble control is reflected in the fact that even though all output measurements of the N systems are available, they do not contain any reference to the N systems that gave rise to these output measurements. Contributions regarding this particular side of the problem include a series of non-pathological sampling results for such sampled observability problems involving anonymized output measurements.

Parts of the results presented here are based on Zeng et al. (2015a,b, 2016a).

3.1 Motivating Example

As a motivating example, we depict in Figure 3.1 an ensemble of three systems evolving in a two-dimensional state space. The initial state and the trajectories of the individual systems are labeled via their shape and color-coding. Suppose that the first coordinates of the three systems are measured at discrete time points, as shown in Figure 3.2. If we had the mentioned labeling of each point in these output measurements (e.g. shape and color-coding), i.e. a mapping between a point in the output measurement to the corresponding system, then we could just treat the observability problem for the ensemble as separate observability problems for the individual systems.

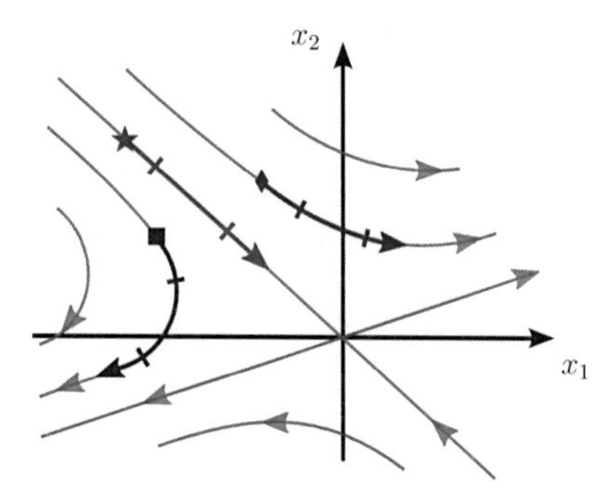

Figure 3.1: Illustration of the discrete ensemble setup. Given a flow, one considers the evolution of an ensemble of particles.

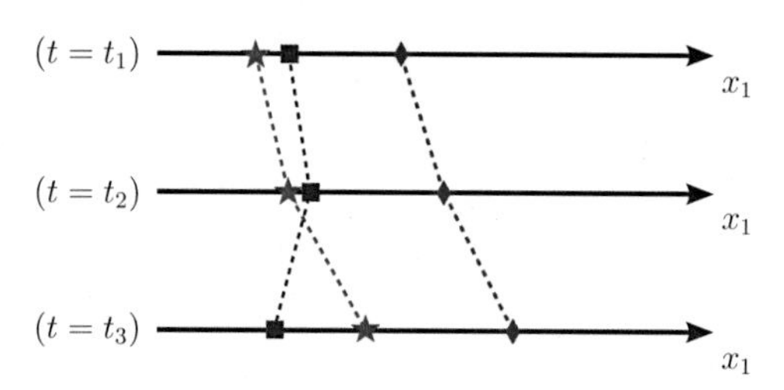

Figure 3.2: Illustration of the output measurements $y = x_1$ of the ensemble at discrete time points with labels (shape and color-coding) for each individual system.

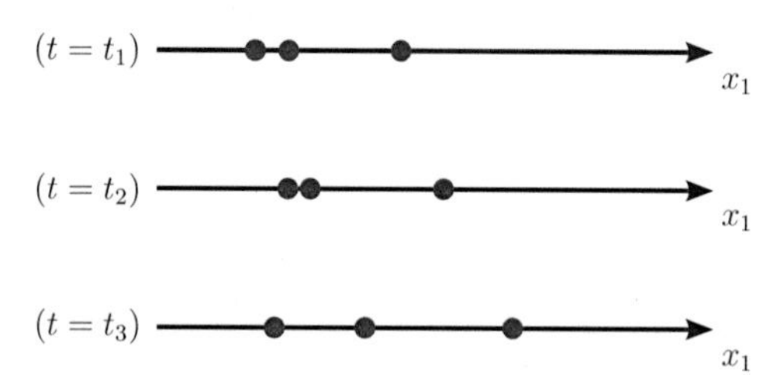

Figure 3.3: Illustration of the output measurements $y = x_1$ at discrete time points without labels for each individual system.

However, in a situation where the measurements are only given in an anonymized way, i.e. any labeling of the output measurements is being "masked", as illustrated in Figure 3.3, the state estimation problem becomes nontrivial. One reason for this is the combinatorial component that is naturally being introduced in this framework, which makes it cumbersome to even formulate the problem.

From a practical point of view, there are several relevant situations in which such abstract anonymization may take place. For example, it could be simply infeasible or too expensive to keep track of output trajectories of individual systems, such as in the related problem of multitarget tracking (Bar-Shalom, 1978; Blom and Bloem, 2000; Kamen, 1992; Smith and Buechler, 1975). Another example which is typical for social populations is that due to privacy issues, the output measurements of individual systems must be treated as "statistics" without reference to the individual system that produced it.

3.2 Problem Setup

As motivated in the beginning of this chapter, we consider an ensemble of N identical linear time-invariant systems, i.e.

$$\begin{aligned} \dot{x}^{(i)}(t) &= Ax^{(i)}(t), \quad x^{(i)}(0) = x_0^{(i)}, \\ y^{(i)}(t) &= Cx^{(i)}(t), \end{aligned} \tag{3.1}$$

where $x^{(i)}(t) \in \mathbb{R}^n$ is the state and $y^{(i)}(t) \in \mathbb{R}^m$ is the output of the ith system, and $i \in \{1, \dots, N\}$. The goal is to reconstruct the set of initial states $x_0^{(i)}$ of the systems in the ensemble. However, unlike in a classical setting, we assume that output measurements of these systems obtained at any discrete time point $t_1, \dots, t_M$ do not contain information $i \mapsto y^{(i)}(t_k)$. The output measurements are thus essentially given in the form of (multi)sets, i.e. for a given time point we obtain

$$Y(t_k) := \{y^{(1)}(t_k), \dots, y^{(N)}(t_k)\},$$

where, in particular, the elements shall be counted with multiplicities when output values of different systems are the same at some point in time.

In such a situation, it is not even straightforward to formulate the sampled observability problem of discrete ensembles in a concise way. It turns out that the measure theoretic framework introduced in the foregoing chapter is natural and useful for this setup as well. Thus, following ideas and insights of the measure theoretic framework in the previous chapter, we model the state of a discrete ensemble via discrete measures. By considering discrete measures, we naturally take the anonymization of the output data into account and, furthermore, we have no particular difficulty when multiple systems in the ensemble produce the exact same output measurement.

In the following, we introduce our notation for discrete measures describing the state of an ensemble and afterwards formulate the state estimation problem in the measure theoretic framework.

Using the notation for the Dirac measure

$$\delta_{x_0^{(i)}}(E) := \begin{cases} 1 & \text{if } x_0^{(i)} \in E, \\ 0 & \text{otherwise,} \end{cases}$$

where $E \subset \mathbb{R}^n$ is a measurable set, we define the initial state of the discrete ensemble via the discrete measure given by

$$\mu_0(E) := \frac{1}{N} \sum_{i=1}^{N} \delta_{x_0^{(i)}}(E).$$

Again, by choosing discrete distributions, each initial state of a system has a "marking", but the markings do not contain individual information.

In view of the problem of reconstructing the initial state distribution μ_0 from output snapshots at discrete times, we observe that the output distribution satisfies the relation

$$\mu_{y(t_k)}(B_y) = \mu_0((Ce^{At_k})^{-1}(B_y)),$$

for any measurable set $B_y \subset \mathbb{R}^m$, which is completely analogous to the case of continuous probability measures considered in the previous chapter. In other words, $\mu_{y(t_k)}$ is simply the pushforward of μ_0 under the mapping $x \mapsto Ce^{At_k}x$.

In Figure 3.4 we illustrate the connection between output distribution and initial state distribution. In this two-dimensional example, the initial state distribution is given by four points. For a particular time instance $t_{k^\star}$ it just so happens that $Ce^{At_{k^\star}}$ is a real projection, i.e. $\|Ce^{At_{k^\star}}\| = 1$. Furthermore, the angle of the projection is such that one particular line of the projection is running through two points, which is, however, easily handled by the formulation of the state of a finite ensemble as a discrete measure.

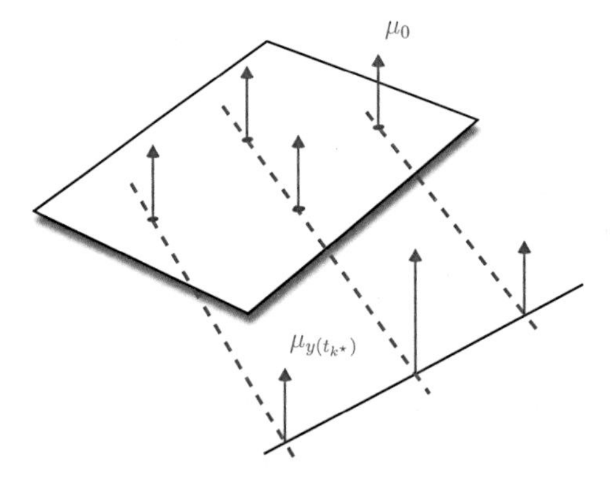

Figure 3.4: This figure illustrates an initial state distribution μ_0 of a discrete ensemble and a particular output snapshot $\mu_{y(t_k^\star)}$ that has a doubled weighting.

Now, given the familiar measure theoretic description of a discrete ensemble through

$$\begin{aligned}\dot{x}(t) &= Ax(t), \quad x(0) \sim \mu_0,\\ y(t) &= Cx(t),\end{aligned}$$

with the discrete measure $\mu_0 = \frac{1}{N}\sum_{i=1}^{N}\delta_{x_0^{(i)}}$, we may rephrase the question of sampled observability of discrete ensembles as follows: under which conditions can we uniquely reconstruct the initial state μ_0 of the ensemble from output snapshots $\mu_{y(t_k)}$ at M discrete times $t_1, \ldots, t_M$?

Similarly to the ensemble observability problem for continuous ensembles, the discrete version treated in this chapter can be viewed as a tomography problem as well. In the following, we illustrate the geometric idea behind this tomography viewpoint in the discrete case. In Figure 3.5, we consider the two cases of a discrete distribution with $N = 1$ and a discrete distribution with $N = 3$ in a two-dimensional state space.

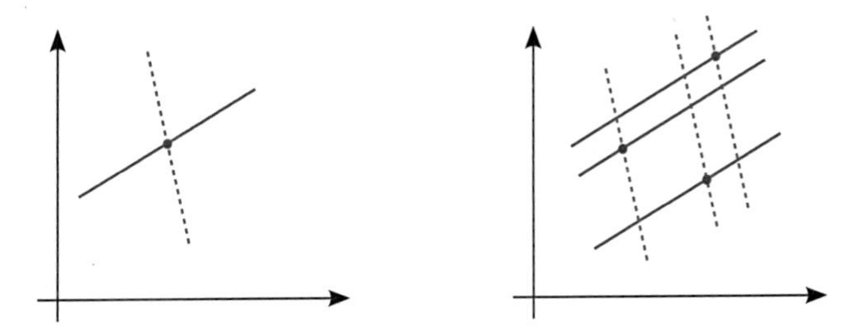

Figure 3.5: Illustration of backprojections $(Ce^{At_k})^{-1}(Y(t_k))$ at two different times for $N = 1$ (left) and $N = 3$ (right), where the dots indicate the initial states. While for the left situation, we are able to uniquely reconstruct the initial state from two output measurements, the situation on the right is different.

We first note that even the classical case $N = 1$ can in fact be considered from a tomography perspective in a meaningful way. From one output at a certain time point t_k we can compute $(Ce^{At_k})^{-1}(\{y(t_k)\})$, which is indicated as a line in this figure; this can be seen as a backprojection (Markoe, 2006). The term "backprojection" stems from the intuitive idea that the projection is "smeared" back to the initial state space along the direction of the projection. Now, one backprojection is clearly not enough for reconstructing the initial state and one thus needs to consider another time point which would, ideally, yield a backprojection at a different angle. The unique intersection then determines the initial state. Now, in the case of more than one system, points where two lines meet are still important. In fact every actual point in the support of μ_0 lies in an intersection of two lines. However, while for $N = 1$ two lines were already sufficient to uniquely determine the initial state, the situation for $N = 3$ is quite different. Therein, two directions are not sufficient to distinguish the actual points from other points which also lie at the intersection of two lines, but which are not in the support of μ_0. The intuitive idea is then to consider more projections at different angles to gradually rule out those points that are in the intersection of M lines but are not in the support of μ_0.

More generally, one may think of the problem of simultaneously solving N systems of M linear equations where for each of the M equations, the N right-hand sides $b_k^{(1)}, \dots, b_k^{(N)}$ are given to us in a randomly shuffled order. Of particular interest in this context are those linear algebraic and geometric properties of the matrix $A \in \mathbb{R}^{M\times n}$ of the corresponding systems of linear equations that would guarantee a unique reconstruction of all solutions $x^{(1)}, \dots, x^{(N)}$ despite the fact that the right-hand sides are given in a randomly shuffled order. We note that, in particular, in such a situation the fact that A has full column rank is no longer sufficient.

3.3 Geometric and Algebraic Characterizations

In this section, we formalize the intuitive ideas of the tomographic approach to the sampled observability problem of discrete ensembles, which yields a first sufficient condition.

Theorem 3.1. *If we are given output snapshots $\mu_{y(t_k)}$ at times $t_1, \dots, t_M$ such that for any set of N non-zero vectors $h^{(1)}, \dots, h^{(N)}$ in $\mathbb{R}^n$ it holds that*

$$\exists k \in \{1, \dots, M\}\, \forall i \in \{1, \dots, N\} \quad h^{(i)} \notin \ker Ce^{At_k}, \tag{3.2}$$

then the initial state μ_0 of the ensemble can be uniquely reconstructed.

Proof. The proof is divided into two steps. In the first step, we show that we can uniquely identify the *support points* of the discrete measure. In a second step we discuss how the individual weights of the support points can be reconstructed.

For the first part, suppose for contradiction that there exists a point $x \in \mathbb{R}^n \setminus \operatorname{supp}(\mu_0)$ which also satisfies $\mu_0(x + \ker Ce^{At_k}) \geq \frac{1}{N}$ for all time points $t_1, \dots, t_M$. By definition of the discrete measure μ_0, we thus have

$$\forall k \in \{1, \dots, M\}\, \exists i \in \{1, \dots, N\} \quad x_0^{(i)} \in x + \ker Ce^{At_k},$$

where $x_0^{(1)}, \dots, x_0^{(N)}$ are actual points in $\operatorname{supp}(\mu_0)$. By virtue of this consideration we find N non-zero vectors $h^{(i)} := x_0^{(i)} - x$ such that

$$\forall k \in \{1, \dots, M\}\, \exists i \in \{1, \dots, N\} \quad h^{(i)} \in \ker Ce^{At_k},$$

which clearly contradicts (3.2). Thus, by virtue of (3.2) and by considering all time points $t_1, \dots, t_M$, any point $x \in \mathbb{R}^n \setminus \operatorname{supp}(\mu_0)$ can indeed also be identified as such.

Regarding the second part, we observe that for each support point $x \in \operatorname{supp}(\mu_0)$ there exists, again due to (3.2), a time point $t_{k(x)}$ at which $Ce^{At_{k(x)}}x^{(i)} = Ce^{At_{k(x)}}x$ only if $x^{(i)} = x$. In other words, for each support point $x \in \operatorname{supp}(\mu_0)$, there exists a time point $t_{k(x)}$ at which

$$\left(Ce^{At_{k(x)}}\right)^{-1}(\{Ce^{At_{k(x)}}x\}) = \{x\},$$

i.e. $Ce^{At_{k(x)}}x \in \operatorname{supp}(\mu_{y(t_{k(x)})})$ can be associated uniquely to $x \in \operatorname{supp}(\mu_0)$. In particular, this yields a constructive way to identify the weights of each support point of μ_0 by

$$\mu_0(x) = \mu_{y(t_{k(x)})}(Ce^{At_{k(x)}}x),$$

which concludes the proof. □

As we have seen from the proof of Theorem 3.1, one can allow for any μ_0 that is generated by an arbitrary choice of $x_0^{(1)}, \ldots, x_0^{(N)} \in \mathbb{R}^n$; in particular the initial states need not be distinct. Considering this very general formulation of a weighted discrete measure certainly has its appeals, such as keeping the connection to the continuous ensemble observability problem particularly close. For simplicity, we will assume in the subsequent presentations that the initial states of the linear systems in the ensemble are pairwise distinct. From a multi-agent systems perspective, for instance, this additional assumption in fact appears to be quite natural.

In this simpler situation, we can give another, perhaps more intuitive, explanation of the result given in Theorem 3.1. Let $x \in \mathbb{R}^n$ with $x \neq x_0^{(i)}$ for all $i \in \{1, \ldots, N\}$. Our goal is to be able to infer this fact from observing only the output snapshots. To this end, consider the following condition that there always exists a time point t_k at which $Ce^{At_k}x$, which is the output value that a system with initial state x would produce at time t_k, does not coincide with any output value of the measured output snapshot at time t_k. The measured output snapshots consist, of course, of the values $Ce^{At_k}x_0^{(i)}$. In mathematical terms, we can thus restate this as the condition

$$\exists k \in \{1, \ldots, M\}\ \forall i \in \{1, \ldots, N\} \quad Ce^{At_k}x \neq Ce^{At_k}x_0^{(i)}.$$

However, since the initial states $x_0^{(i)}$ with $i \in \{1, \ldots, N\}$ are undisclosed to us, we cannot possibly guarantee this precise condition, but we can guarantee a more conservative version, i.e. that

$$\exists k \in \{1, \ldots, M\}\ \forall i \in \{1, \ldots, N\} \quad Ce^{At_k}(x - x_0^{(i)}) \neq 0$$

for all $x, x_0 \in \mathbb{R}^n$ such that $x \neq x_0^{(i)}, i \in \{1, \ldots, N\}$. This discussion also yields (3.2).

A Richness Property for a Family of Subspaces

In this subsection, we further highlight the property given in Theorem 3.1. The crucial condition (3.2) may be geometrically viewed as a richness of the family of subspaces, as the subspaces within the family have to exhibit "sufficient movement in all dimensions". In view of this analogy, we introduce the following slightly more encompassing definition.

Definition 3.2 (Richness of a family of subspaces). Let K be an arbitrary index set. We say that a family of subspaces $(W_k)_{k\in K}$ is *rich of order* N if for any set of N non-zero vectors $h^{(1)}, \ldots, h^{(N)} \in \mathbb{R}^n$ we have

$$\exists k \in K \ \ \forall i \in \{1, \ldots, N\} \quad h^{(i)} \notin W_k.$$

In the context of sampled observability of discrete ensembles, the family of subspaces that we are interested in is $W_k := \ker Ce^{At_k}$. In this situation, our ultimate goal is thus to find conditions on the linear system (A, C), as well as the sampling times $t_1, \ldots, t_M$, so that the corresponding family $(W_k)_{k\in\{1,\ldots,M\}}$ is rich of order N.

To further illustrate the concept of richness of a family of subspaces, we consider the following example.

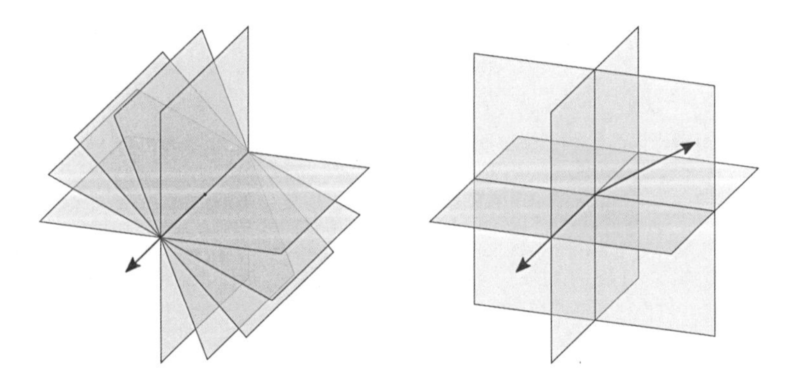

Figure 3.6: The family of subspaces on the left is not rich of order 1. The family of subspaces on the right is rich of order 1 but not rich of order 2.

Example 3.3. Consider the two families of subspaces depicted in Figure 3.6. The left picture in Figure 3.6 shows a family of subspaces that is not rich of order 1. Every subspace contains the highlighted direction. Note also how one can describe this family simply in terms of a rotation about this highlighted direction. The picture on the right shows a family that is rich of order 1. One cannot find a non-zero vector which is contained in all three subspaces, which is nothing but a geometric way of expressing that the intersection of the subspaces is trivial. However, as hinted by the depicted choice of directions, the family is not rich of order 2.

It is quite remarkable that our initial question under which conditions the family of linear mappings $(Ce^{At_k})_{k\in\{1,\dots,M\}}$ *separates* discrete distributions with N support points, i.e. sets of N points, can in fact be recast and illustrated as a geometric separation of sets of N unit vectors by the subspaces $W_k = \ker Ce^{At_k}$, cf. Figure 3.6.

Lastly, we note that it can be verified that the family of subspaces in the left picture in Figure 3.6 can be generated via $W_k = \ker Ce^{At_k}$ by the unobservable system

$$\dot{x}(t) = \left(\begin{array}{cc|c} -1 & 1 & 0 \\ 0 & 0 & 0 \\ \hline 0 & 0 & 0 \end{array}\right) x(t), \quad y(t) = \left(\begin{array}{cc|c} 1 & 0 & 0 \end{array}\right) x(t),$$

given in a Kalman decomposition form. Note that the observable part of this system is the familiar system (2.5) that we had introduced in Example 2.4 in the previous chapter.

It is instructive to discuss this richness property in the case that the output is measured continuously in time. For an observable system one would expect that it should be in principle possible to obtain a family of subspaces of the form $\ker Ce^{At}$ that is rich of a fixed order $N \geq 1$. The following result readily confirms this.

Proposition 3.4. *Let (A, C) be an observable system. Then for any given (at most) countable family of non-zero vectors $h^{(1)}, h^{(2)}, \dots$ there exists $t \geq 0$ such that*

$$\forall i \in \{1, 2, \dots\} \quad h^{(i)} \notin \ker Ce^{At}.$$

Proof. We show the contraposition of the claim. For this purpose, we suppose that there exists a family of non-zero vectors $h^{(1)}, h^{(2)}, \dots$ such that

$$\forall t \geq 0 \; \exists i \in \{1, 2, \dots\} \quad h^{(i)} \in \ker Ce^{At}.$$

Then, defining $T_i = \{t \geq 0 \, : \, Ce^{At}h^{(i)} = 0\}$ we have $\bigcup_{i \in \{1,2,\dots\}} T_i = [0, \infty)$. This, however, contradicts that the signal $t \mapsto Ce^{At}h^{(i)}$ has isolated zeros due to observability of (A, C) and $h^{(i)} \neq 0$. □

So far, we have provided a first geometric condition for the sampled observability problem for discrete linear ensembles to be solvable. This sufficient condition requires that the family of subspaces $\ker Ce^{At_k}$, parameterized by $k \in \{1, \dots, M\}$, is rich of order N, where N is the number of systems in the considered ensemble. In the following, we reformulate this condition into a form which displays a close connection to the tomography-based approach of the previous chapter.

To this end, we relate the condition that the family $(\ker Ce^{At_k})_{k \in \{1,\dots,M\}}$ is rich of order N to a condition formulated in terms of algebraic varieties.

Theorem 3.5. *The family of subspaces $(\ker Ce^{At_k})_{k \in \{1,\dots,M\}}$ is rich of order N if and only if the union $\bigcup_{k \in \{1,\dots,M\}} \operatorname{im}(Ce^{At_k})^\top$ is contained in no algebraic variety defined by*

$$\langle x, h^{(1)} \rangle \cdots \langle x, h^{(N)} \rangle = 0,$$

with some non-zero $h^{(1)}, \dots, h^{(N)} \in \mathbb{R}^n$.

Proof. By definition, the family $(\ker Ce^{At_k})_{k \in \{1,\dots,M\}}$ is not rich of order N if there exist non-zero vectors $h^{(1)}, \dots, h^{(N)} \in \mathbb{R}^n$ so that

$$\forall k \in \{1, \dots, M\} \; \exists i \in \{1, \dots, N\} \quad Ce^{At_k}h^{(i)} = 0. \tag{3.3}$$

This is equivalent to the existence of non-zero vectors $h^{(1)}, \dots, h^{(N)} \in \mathbb{R}^n$ so that

$$\forall k \in \{1, \dots, M\} \; \forall v \in \mathbb{R}^m \quad \langle v, Ce^{At_k}h^{(1)} \rangle \cdots \langle v, Ce^{At_k}h^{(N)} \rangle = 0, \tag{3.4}$$

which is, in turn, equivalent to $\bigcup_{k \in \{1,\dots,M\}} \operatorname{im}(Ce^{At_k})^\top$ being contained in the algebraic variety defined by $\langle x, h^{(1)} \rangle \cdots \langle x, h^{(N)} \rangle = 0$. While (3.3) trivially implies (3.4), to see the converse, suppose for contradiction that there exist non-zero $h^{(1)}, \dots, h^{(N)} \in \mathbb{R}^n$ so that (3.4) holds while (3.3) does not. Thus,

$$\exists k \in \{1, \dots, M\} \; \forall i \in \{1, \dots, N\} \quad Ce^{At_k}h^{(i)} \neq 0,$$

which contradicts (3.4), since the finite union of the subspaces $\ker \langle Ce^{At_k}h^{(i)}, \cdot \rangle$, which have co-dimension 1, cannot be the whole $\mathbb{R}^n$. □

The last argument in the proof of the theorem can be summarized as follows.

Proposition 3.6. *The family* $(\ker \langle v, \cdot \rangle)_{v \in \mathbb{R}^n}$ *is rich of any order* N, *i.e. for any finite number of non-zero vectors* $h^{(1)}, \dots, h^{(N)} \in \mathbb{R}^n$, *it holds that*

$$\exists v \in \mathbb{R}^m \; \forall i \in \{1, \dots, N\} \quad \langle v, h^{(i)} \rangle \neq 0.$$

This shows the following result, which may be regarded as a (perhaps trivial) discrete counterpart to the Cramér-Wold theorem.

Corollary 3.7. *A discrete measure with finitely many support points in* $\mathbb{R}^n$ *is uniquely determined by the family of all its pushforward measures along* $x \mapsto \langle v, x \rangle$, $v \in \mathbb{R}^n$.

This result will play an important role in the next section, where we introduce a moment-based framework for the discrete situation.

An Algebraic Characterization

Before we proceed with introducing the aforementioned moment-based framework, in this subsection, we derive a linear algebraic characterization of the richness property as opposed to the fully geometric characterization given so far. The following lemma characterizes a family of subspaces which is *not* rich of order N.

Lemma 3.8. *For a family of subspaces* $W_1, \dots, W_M$ *there exists a family of* N *non-zero vectors* $h^{(1)}, \dots, h^{(N)} \in \mathbb{R}^n$ *such that*

$$\forall k \in \{1, \dots, M\} \; \exists i \in \{1, \dots, N\} \quad h^{(i)} \in W_k, \tag{3.5}$$

if and only if there exists a partition $P_1, \dots, P_N$ *of the set* $\{1, \dots, M\}$ *such that*

$$\forall i \in \{1, \dots, N\} \quad \bigcap_{k \in P_i} W_k \quad \textit{nontrivial.}$$

Proof. The property in (3.5) naturally induces a partition of $\{1, \dots, M\}$ into $\tilde{N} \leq N$ index subsets $P_{i_1}, \dots, P_{i_{\tilde{N}}}$ such that

$$\forall \ell \in \{1, \dots, \tilde{N}\} \quad \operatorname{span}(\{h^{(i_\ell)}\}) \subset \bigcap_{k \in P_{i_\ell}} W_k,$$

i.e. the one-dimensional subspace $\operatorname{span}(\{h^{(i_\ell)}\})$ is contained in the subspace $\bigcap_{k \in P_{i_\ell}} W_k$. Therefore, the intersection $\bigcap_{k \in P_{i_\ell}} W_k$ is nontrivial. One can now arbitrarily divide this partition further such that one ends up with N index subsets. The intersections (since now taken over a possibly smaller index subset) will remain nontrivial.

To show the converse, pick out of each (nontrivial) intersection $\bigcap_{k \in P_i} W_k$ a non-zero vector $h^{(j)}$ such that one ends up with a family of non-zero $h^{(1)}, \dots, h^{(\tilde{N})}$ that falsifies richness of order $\tilde{N} \leq N$ and thus richness of order N of the family $W_1, \dots, W_M$. □

Combining Lemma 3.8 with Theorem 3.1, we finally arrive at the following sufficient condition for sampled observability of a discrete ensemble.

Theorem 3.9. *If we are given output snapshots $\mu_{y(t_k)}$ of a discrete ensemble at times $t_1, \dots, t_M$ such that for any partition of $\{1, \dots, M\}$ into index subsets $P_1, \dots, P_N$*

$$\exists i \in \{1, \dots, N\} \quad \bigcap_{k \in P_i} \ker Ce^{At_k} \quad \textit{trivial} \tag{3.6}$$

then the initial state μ_0 of the ensemble can be uniquely reconstructed.

We note that the property in Theorem 3.9 can also be restated as follows. For any partition of $\{1, \dots, M\}$ into index subsets $P_1, \dots, P_N$, at least one of the matrices of the form $(Ce^{At_k})_{k \in P_i}$ needs to have full column rank.

It is instructive to discuss how Theorem 3.9 generalizes the classical case of $N = 1$, i.e. the classical sampled observability problem for linear systems. In the language of Theorem 3.9, we need to consider all the intersections (3.6) over all partitions $P_1, \dots, P_N$. For $N = 1$ there is of course only one partition to consider, namely the set $\{1, \dots, M\}$ itself. Therefore, only the intersection of $\ker Ce^{At_k}$ over all indices $k \in \{1, \dots, M\}$ needs to be trivial, which is the claimed classical result.

Before we attempt to build upon Theorem 3.9 to arrive at a more explicit sufficient condition for sampled observability of discrete ensembles, we will discuss the applicability of moment-based approaches, as introduced in Chapter 2, to the discrete case considered in this chapter. This is indeed motivated by the moment-based approach to the continuous ensemble observability problem. Moreover, it is also interesting to see how the mathematical structures that we encountered so far are translated in a moment-based, i.e. polynomial, approach.

3.4 Moment Dynamics of Discrete Ensembles

In this section, we introduce a moment-based approach for the study of the considered discrete ensembles in this chapter. We first establish different polynomial descriptions of discrete measures with finitely many support points, which can be viewed as a finite analogue of moments of continuous distributions. Afterwards we show that, similarly as in the continuous ensemble observability problem, the moment dynamics are described in terms of tensor systems.

Definition of Moments of Discrete Ensembles

For any discrete measure μ defined on $\mathbb{R}^n$, with support points $x^{(1)}, \dots, x^{(N)}$, we may define the moments of μ by

$$\mathbb{E}[x_1^{\alpha_1} \cdots x_n^{\alpha_n}] = \int_{\mathbb{R}^n} x_1^{\alpha_1} \cdots x_n^{\alpha_n} \, \mathrm{d}\mu = \sum_{i=1}^{N} (x^{(i)})^{\alpha}, \tag{3.7}$$

similarly as in the case of continuous probability distributions. We expect that μ is uniquely characterized by its moments, though we would also expect that to this end a *finite* number of moments suffices. In the following, we discuss how one can extract a finite number of symmetric statistics of μ that determine μ uniquely.

Consider first the simpler case of a discrete measure μ_y defined on $\mathbb{R}$ with scalar real values $y^{(1)}, \dots, y^{(N)}$ as support points. The key idea for extracting only the relevant moments of a discrete measure, i.e. a minimal number of moments which determine μ_y uniquely, is to associate to the discrete measure μ_y the polynomial

$$p(s) := (s - y^{(1)}) \cdots (s - y^{(N)}).$$

We notice that, by construction, this polynomial vanishes at the support points, or, in other words, the support points of the measure μ_y are encoded in the zeros of p. As for the zeros of p, these do not have an explicit canonical numbering, so that the anonymization within the ensemble is indeed preserved within this reformulation. More generally, the representation using the polynomial p is a useful equivalent description for a (scalar) discrete measure μ_y.

If we expand the linear factorization of p, we obtain

$$p(s) = \sum_{p=0}^{N} (-1)^p e_p(y^{(1)}, \dots, y^{(N)}) s^{N-p},$$

where $e_0 \equiv 1$ trivially since p is monic, and $e_1, \dots, e_N$ are known as the elementary symmetric polynomials (cf. *Vieta's formulas*). An explicit representation of the elementary symmetric polynomials is given by

$$e_p(y^{(1)}, \dots, y^{(N)}) = \sum_{1 \leq j_1 < \cdots < j_p \leq N} y^{(j_1)} \cdots y^{(j_p)}.$$

In the following example, we provide an illustration of the general idea for a simple case in which we consider a discrete measure with three support points.

Example 3.10. Consider a discrete measure μ_y with scalar support points $y^{(1)}, y^{(2)}, y^{(3)}$. As discussed for the general case, we associate to the discrete measure μ_y the polynomial $p(s) = (s - y^{(1)})(s - y^{(2)})(s - y^{(3)})$. Expanding this polynomial yields

$$p(s) = s^3 - e_1(y^{(1)}, y^{(2)}, y^{(3)}) s^2 + e_2(y^{(1)}, y^{(2)}, y^{(3)}) s - e_3(y^{(1)}, y^{(2)}, y^{(3)}),$$

where the coefficients are given by

$$\begin{aligned} e_1(y^{(1)}, y^{(2)}, y^{(3)}) &= y^{(1)} + y^{(2)} + y^{(3)}, \\ e_2(y^{(1)}, y^{(2)}, y^{(3)}) &= y^{(1)} y^{(2)} + y^{(1)} y^{(3)} + y^{(2)} y^{(3)}, \\ e_3(y^{(1)}, y^{(2)}, y^{(3)}) &= y^{(1)} y^{(2)} y^{(3)}. \end{aligned} \tag{3.8}$$

We note that the coefficients are in fact the symmetric measurement equations (1.2) mentioned in the introduction. Since a polynomial is uniquely characterized by its coefficients, we conclude that the symmetric equations (3.8) of the values $y^{(1)}, y^{(2)}, y^{(3)}$ characterize the distribution uniquely. Intuitively, it should not come as a surprise that the equations characterizing μ_y are symmetric in the sense that any permutation of the variables yields the same value. This is in perfect accordance with the fact that for a discrete distribution, there is no canonical labeling of the support points.

Based on this intuitive point of view, approaches for the related problem of multitarget tracking employing symmetric polynomials have been attracting interest already since 1990, see Kamen (1992); Kamen and Sastry (1993). Therein, the starting point is indeed the intuitive idea to "convert the measurement data (with association not known) into a measurement equation by defining new measurements that are symmetric functionals of the original measurements" (Kamen and Sastry, 1993), which is typically done by (heuristically) introducing power sum symmetric polynomials

$$p_p(y^{(1)}, \dots, y^{(N)}) = (y^{(1)})^p + \dots + (y^{(N)})^p.$$

In Kamen and Sastry (1993), it is noted that employing sums of products such as (3.8) yields a better performance of practical filters, which is why one then confines oneselves to these for practical implementations. With the point of view of generating functions for the discrete distributions taken in this section, the choice for (3.8) seems more systematic. In fact, the connection to the sum of powers equation is also straightforward from the present point of view: by virtue of *Newton's identities* in the univariate case, there is a one-to-one relation between power sum symmetric polynomials and elementary symmetric polynomials given by

$$pe_p(y^{(1)}, \dots, y^{(N)}) = \sum_{i=1}^{p} (-1)^{i-1} e_{p-i}(y^{(1)}, \dots, y^{(N)}) p_i(y^{(1)}, \dots, y^{(N)}),$$

so that eventually, every elementary symmetric polynomial is uniquely characterized by sums of powers of its zeros. Since the power sum symmetric polynomials are in fact nothing but moments of scalar discrete measures by the classical definition (3.7), this demonstrates that in the scalar case, the first N moments characterize a discrete measure with N support points uniquely. We note that (3.8), on the other hand, are strictly speaking no moments by the usual definition. But as the above discussion reveals, they are an equivalent description analogous to classical moments, which is why we simply refer to these, and their multivariate generalizations, as moments for convenience.

In the following, we describe how one can also define these moments for multivariate discrete measures, which is also of interest in its own right. We will restrict our attention to symmetric measurement equations that originate from elementary symmetric polynomials as the analogous derivation for power sum symmetric polynomials is straightforward, partly due to the fact that they are moments by the classical definition. Given a discrete measure μ_x with support points $x^{(1)}, \dots, x^{(N)} \in \mathbb{R}^n$ and a single-output $y = \langle c, x \rangle$ with a vector $c \in \mathbb{R}^n$, we associate to the resulting output distribution μ_y the polynomial

$$p_y(s) := (s - \langle c, x^{(1)} \rangle) \cdots (s - \langle c, x^{(N)} \rangle).$$

We then consider the coefficients of p_y in order to extract a characterization of the support points $x^{(1)}, \dots, x^{(N)}$ in terms of symmetric polynomial equations of these. This approach may be described as the reduction of a multivariate problem, of which the solution does not appear straightforward, to a simpler univariate one by considering all one-dimensional projections. It is therefore very much in the spirit of the Cramér-Wold theorem, and has in fact been one of the key ideas throughout this thesis.

Example 3.11. Consider the case that $n = 2$ and $N = 2$. A direct computation yields

$$e_1(\langle c, x^{(1)}\rangle, \langle c, x^{(2)}\rangle) = \langle c, x^{(1)}\rangle + \langle c, x^{(2)}\rangle = \begin{pmatrix} c_1 & c_2 \end{pmatrix} \begin{pmatrix} x_1^{(1)} + x_1^{(2)} \\ x_2^{(1)} + x_2^{(2)} \end{pmatrix}$$

for the first elementary symmetric polynomial. Similarly, we have

$$e_2(\langle c, x^{(1)}\rangle, \langle c, x^{(2)}\rangle) = \langle c, x^{(1)}\rangle\langle c, x^{(2)}\rangle = \begin{pmatrix} c_1^2 & \sqrt{2}c_1c_2 & c_2^2 \end{pmatrix} \begin{pmatrix} x_1^{(1)}x_1^{(2)} \\ \frac{1}{\sqrt{2}}(x_1^{(1)}x_2^{(2)} + x_1^{(2)}x_2^{(1)}) \\ x_2^{(1)}x_2^{(2)} \end{pmatrix}$$

for the second elementary symmetric polynomial. We recognize the occurring matrix $(c_1^2 \ \sqrt{2}c_1c_2 \ c_2^2)$ as the tensor power $C^{[2]}$ that we introduced in the foregoing chapter, cf. Example 2.13. Moreover, it is intuitively clear that the five symmetric equations in total for the support points must determine the discrete measure μ_x uniquely. More generally, we see that $e_p(\langle c, x^{(1)}\rangle, \dots, \langle c, x^{(N)}\rangle)$ can always be split into a product of $C^{[p]}$ and a symmetric characterization for μ_x, which we will denote as $x^{\langle p\rangle}$, so that

$$e_p(\langle c, x^{(1)}\rangle, \dots, \langle c, x^{(N)}\rangle) =: C^{[p]}x^{\langle p\rangle}.$$

We note that a convenient explicit description of $x^{\langle p\rangle}$ is not easy to find in general, hence the implicit definition of the vector $x^{\langle p\rangle}$.

Furthermore, for this example with $n = 2$ and $N = 2$, a direct computation yields

$$\begin{aligned} \frac{d}{dt}(x_1^{(1)}x_1^{(2)}) &= 2a_{11}x_1^{(1)}x_1^{(2)} + a_{12}(x_1^{(1)}x_2^{(2)} + x_1^{(2)}x_2^{(1)}), \\ \frac{d}{dt}\frac{1}{\sqrt{2}}(x_1^{(i)}x_2^{(j)}) &= \frac{1}{\sqrt{2}}(a_{21}x_1^{(i)}x_1^{(j)} + (a_{11} + a_{22})x_1^{(i)}x_2^{(j)} + a_{12}x_2^{(i)}x_2^{(j)}), \\ \frac{d}{dt}(x_2^{(1)}x_2^{(2)}) &= 2a_{22}x_2^{(1)}x_2^{(2)} + a_{21}(x_1^{(1)}x_2^{(2)} + x_1^{(2)}x_2^{(1)}). \end{aligned}$$

Summing up the terms in the second equation to obtain the second component of the vector $x^{\langle 2\rangle}$, we arrive at the representation of the dynamics of $x^{\langle 2\rangle}$ given by

$$\frac{d}{dt}x^{\langle 2\rangle} = \begin{pmatrix} 2a_{11} & \sqrt{2}a_{12} & 0 \\ \sqrt{2}a_{21} & a_{11} + a_{22} & \sqrt{2}a_{12} \\ 0 & \sqrt{2}a_{21} & 2a_{22} \end{pmatrix} x^{\langle 2\rangle}.$$

Therein, the system matrix is recognized as $A_{[2]}$ from Example 2.13 in the previous chapter. Therefore, we can conclude for this example that the second order moments of the discrete distribution μ_y and the second order moments of the discrete distribution μ_x are related by

$$\begin{aligned} \frac{d}{dt}x^{\langle 2\rangle} &= A_{[2]}x^{\langle 2\rangle}, \\ y^{\langle 2\rangle} &= C^{[2]}x^{\langle 2\rangle}. \end{aligned}$$

Again, these are the familiar tensor systems that we introduced in Chapter 2 as part of the development of the moment-based framework for studying (continuous) ensemble observability of dynamical systems.

In the remainder of this section, we generalize the foregoing approach to allow for the consideration of multivariate outputs. We also show that, analogously to the case of absolutely continuous distributions, the moments of a discrete output distribution and the moments of the corresponding state distribution are related by tensor systems.

Given support points $y^{(1)}, \dots, y^{(N)}$ in $\mathbb{R}^m$, the key idea is to consider for an arbitrary fixed $v \in \mathbb{R}^m$, the polynomial

$$p_v(s) := (s - \langle v, y^{(1)} \rangle) \cdots (s - \langle v, y^{(N)} \rangle),$$

i.e. to reduce the multivariate problem to univariate problems. We know from Corollary 3.7 that the family of all one-dimensional projections $\mu_{\langle v,y \rangle}$, where $v \in \mathbb{R}^n$, of the original measure μ_y determine μ_y uniquely. Expanding the polynomial $p_v(s)$ yields

$$p_v(s) = \sum_{p=0}^{N} (-1)^p e_p(\langle v, y^{(1)} \rangle, \dots, \langle v, y^{(N)} \rangle) s^{N-p}.$$

Now, similarly to the single-output case, since the mapping

$$v \mapsto e_p(\langle v, y^{(1)} \rangle, \dots, \langle v, y^{(N)} \rangle)$$

is a homogeneous polynomial of degree p, we can write

$$e_p(\langle v, y^{(1)} \rangle, \dots, \langle v, y^{(N)} \rangle) =: \langle v^{[p]}, y^{\langle p \rangle} \rangle,$$

from which we can extract symmetric polynomial equations $y^{\langle p \rangle}$ for the multivariate output measurements. Furthermore these define the discrete distribution uniquely, as established by the following theorem.

Theorem 3.12. *Let $\mu_y = \frac{1}{N} \sum_{i=1}^{N} \delta_{y^{(i)}}$ be a discrete measure with support points in $\mathbb{R}^m$. Then the symmetric measurement equations $y^{\langle p \rangle}$ for $p \in \{1, \dots, N\}$ defined by*

$$\forall v \in \mathbb{R}^m \quad e_p(\langle v, y^{(1)} \rangle, \dots, \langle v, y^{(N)} \rangle) = \langle v^{[p]}, y^{\langle p \rangle} \rangle,$$

determine μ_y uniquely.

Proof. We first show that $y^{\langle p \rangle}$ is in fact well-defined. The mapping

$$\tilde{\psi}_p : \mathbb{R}^N \to \mathbb{R}, \quad (\tilde{y}^{(1)}, \dots, \tilde{y}^{(N)}) \mapsto e_p(\tilde{y}^{(1)}, \dots, \tilde{y}^{(N)})$$

is a homogeneous polynomial of degree p. Thus, if we consider the composition

$$v \mapsto \tilde{\psi}_p(\langle v, y^{(1)} \rangle, \dots, \langle v, y^{(N)} \rangle),$$

then this is also a polynomial of degree p. Hence there exists a unique vector $y^{\langle p \rangle}$ of coefficients, so that for all $v \in \mathbb{R}^m$

$$\langle v^{[p]}, y^{\langle p \rangle} \rangle = \tilde{\psi}_p(\langle v, y^{(1)} \rangle, \dots, \langle v, y^{(N)} \rangle) = e_p(\langle v, y^{(1)} \rangle, \dots, \langle v, y^{(N)} \rangle),$$

which shows that $y^{\langle p \rangle}$ is indeed well-defined.

The last step is to show that the symmetric equations $y^{\langle p\rangle}$ for $p \in \{1,\dots,N\}$ determine the underlying discrete measure uniquely. Let μ_y' and μ_y'' be two discrete measures such that all their symmetric measurement equations $y'^{\langle p\rangle} = y''^{\langle p\rangle}$ are equal for all $p \in \{1,\dots,N\}$. By definition of the symmetric measurement equations, this implies that

$$\forall v \in \mathbb{R}^m \quad e_p(\langle v, y'^{(1)}\rangle, \dots, \langle v, y'^{(N)}\rangle) = e_p(\langle v, y''^{(1)}\rangle, \dots, \langle v, y''^{(N)}\rangle).$$

Since in the univariate case, moment-determinacy is trivially fulfilled, this means that all projections of the distributions along $v \in \mathbb{R}^m$ agree. By virtue of Corollary 3.7 we conclude that $\mu_y' = \mu_y''$. □

Recall that in the single-output case we have, e.g., $y^{\langle p\rangle} = e_p(y^{(1)}, \dots, y^{(N)})$, cf. (3.8). Measurement equations for the two-dimensional case are given in the previous subsection. To conclude, this provides a systematic approach to define and also obtain relevant symmetric measurement equations for arbitrary dimensions of the output. Next, we show that the dynamics of the moments $y^{\langle p\rangle}$ are described by linear tensor systems, as we have already seen for the specific example in the previous subsection.

Dynamics of Moments of Discrete Ensembles

Based on the general approach for multivariate output (and state) measurements, the following theorem describes the dynamics of the moments in the general case.

Theorem 3.13. *Consider a discrete ensemble evolving under a linear system (A, C). The dynamics of the moments $x^{\langle p\rangle}$ and $y^{\langle p\rangle}$ are governed by*

$$\begin{aligned} \frac{d}{dt} x^{\langle p\rangle} &= A_{[p]} x^{\langle p\rangle}, \qquad x^{\langle p\rangle}(0) = x_0^{\langle p\rangle}, \\ y^{\langle p\rangle} &= C^{[p]} x^{\langle p\rangle}. \end{aligned}$$

Proof. Since $y^{(i)} = Ce^{At}x_0^{(i)}$, we have

$$\langle v, y^{(i)}\rangle = \langle v, Ce^{At}x_0^{(i)}\rangle = \langle (Ce^{At})^\top v, x_0^{(i)}\rangle,$$

and thus

$$e_p(\langle v, y^{(1)}\rangle, \dots, \langle v, y^{(N)}\rangle) = e_p(\langle (Ce^{At})^\top v, x_0^{(1)}\rangle, \dots, \langle (Ce^{At})^\top v, x_0^{(N)}\rangle).$$

By the definition of $x^{\langle p\rangle}$ in Theorem 3.12 and the fact that $(\tilde{A}^\top)^{[p]} = (\tilde{A}^{[p]})^\top$, we obtain

$$e_p(\langle (Ce^{At})^\top v, x_0^{(1)}\rangle, \dots, \langle (Ce^{At})^\top v, x_0^{(N)}\rangle) = \langle ((Ce^{At})^\top v)^{[p]}, x_0^{\langle p\rangle}\rangle = \langle v^{[p]}, (Ce^{At})^{[p]} x_0^{\langle p\rangle}\rangle.$$

Therefore, we arrive at

$$\forall v \in \mathbb{R}^m \quad \langle v^{[p]}, y^{\langle p\rangle}\rangle = \langle v^{[p]}, (Ce^{At})^{[p]} x_0^{\langle p\rangle}\rangle,$$

and eventually $y^{\langle p\rangle} = (Ce^{At})^{[p]} x_0^{\langle p\rangle} = C^{[p]} e^{A_{[p]}t} x_0^{\langle p\rangle}$. □

To conclude, Theorem 3.12 shows how to define symmetric measurement equations for arbitrary discrete measures and that these contain all the relevant information of the discrete measures. Moreover, with Theorem 3.13 we managed to show that the evolution of the finite number of relevant moments of a discrete measure is governed by linear tensor systems, i.e.,

$$\begin{aligned}
\frac{d}{dt}\begin{pmatrix} x^{\langle 1 \rangle} \\ \vdots \\ x^{\langle N \rangle} \end{pmatrix} &= \begin{pmatrix} A_{[1]} & & \\ & \ddots & \\ & & A_{[N]} \end{pmatrix} \begin{pmatrix} x^{\langle 1 \rangle} \\ \vdots \\ x^{\langle N \rangle} \end{pmatrix}, \\
\begin{pmatrix} y^{\langle 1 \rangle} \\ \vdots \\ y^{\langle N \rangle} \end{pmatrix} &= \begin{pmatrix} C^{[1]} & & \\ & \ddots & \\ & & C^{[N]} \end{pmatrix} \begin{pmatrix} y^{\langle 1 \rangle} \\ \vdots \\ y^{\langle N \rangle} \end{pmatrix}.
\end{aligned} \tag{3.9}$$

Therefore, in view of the sampled observability of discrete ensembles, we can directly conclude at this point that if the first N tensor systems $(A_{[p]}, C^{[p]}), p \in \{1, \dots, N\}$, are observable, then one can choose M such that the sampling frequency is above the critical frequency of the fastest of those tensor systems $(A_{[p]}, C^{[p]})$. Again, the crucial fact is that in the discrete ensemble observability problem considered here, only a finite number of tensor systems occur.

In the next section, we reconsider the combinatorial characterization of sampled observability of discrete ensembles and show how for the specific class of single-output systems with distinct real eigenvalues, observability of (A, C) alone guarantees sampled observability of a discrete ensemble. The scope of this result is then extended to arbitrary observable systems. More precisely, we show that if (A, C) is observable, then in order to have sampled observability for the discrete ensemble consisting of N copies of a linear system (A, C), the sampling frequency must be chosen N times higher than the critical frequency of the linear system (A, C). It is perhaps surprising that even though we obtained the formulation (3.9) for the moments of the discrete ensemble in terms of the first N tensor systems, we eventually learn that only the observability of (A, C) is relevant for the sampled observability of discrete linear ensembles. The explanation of this seemingly contradicting circumstance is that the vectors $\left(x^{\langle p \rangle}\right)_{p \in \{1,\dots,N\}}$, i.e. the relevant moments of the discrete measure that we would like to distinguish, cannot be arbitrary, but are in fact parameterized via the polynomial mapping

$$\psi : [(x^{(1)}, \dots, x^{(N)})] \mapsto \begin{pmatrix} x^{\langle 1 \rangle} \\ \vdots \\ x^{\langle N \rangle} \end{pmatrix},$$

where $[(x^{(1)}, \dots, x^{(N)})]$ denotes an element of the quotient space $(\mathbb{R}^n \times \cdots \times \mathbb{R}^n)/\mathfrak{S}_N$, and $\mathfrak{S}_N$ is the symmetric group over $\{1, \dots, N\}$. We note that the description through an element of this quotient space is equivalent to the description using discrete measures and that we adopted this particular notation here for the sole purpose of highlighting the aforementioned parametrization via the support points of the discrete distribution.

3.5 Non-Pathological Sampling Results

The majority of the foregoing part of this chapter was concerned with studying and emphasizing the connection of the sampled observability problem of discrete ensembles to the conceptually more general ensemble observability problem presented in the previous chapter. In particular, we obtained a first result that characterized sampled observability of a discrete ensemble in terms of observability of the first N tensor systems. In this section, we first show that a sampled-data theoretic approach to the sampled observability problem of discrete ensembles leads to a sharper sufficient condition. In contrast to the result obtained by the moment-based framework, the sufficient condition derived from a sampled-data theoretic perspective requires only the observability of (A, C), and, in particular, not the observability of its higher order tensor systems. The sampled-data theoretic approach is first illustrated by the familiar class of single-output systems with distinct real eigenvalues. In fact, in this case it is straightforward to build upon the combinatorial characterization given by Theorem 3.9 of this chapter. We then discuss the generalization of this approach to arbitrary observable systems (A, C). This generalization is based on a non-pathological sampling result for irregularly sampled linear systems, which is briefly reviewed. The main theorem for sampled observability of an ensemble of N linear systems (A, C) is then the result that we precisely need N times the number of sampling times that we would need for the non-pathological sampling of one single linear system (A, C).

The second part of this section is concerned with exploiting the sampled-data theoretic approach to obtain results for more general cases of discrete linear ensembles, such as ensembles of heterogeneous linear systems and ensembles of interconnected linear systems.

Sampled Observability of Discrete Ensembles

With Theorem 3.9 we obtained a first sufficient condition for the sampled observability of discrete ensembles, yet the condition is implicit as it only states a condition on the family of linear mappings Ce^{At_k}, the verification of which is quite impractical. In the following, we show how Theorem 3.9, nevertheless, can serve as the basis for formulating a more explicit sufficient condition. In particular, we combine the geometric ideas of the previous section with a (classical) sampled-data control theoretic framework to obtain the main result.

We begin by slightly simplifying the formulation of Theorem 3.9. We note that by choosing the number of measurement instances sufficiently large, we can achieve that for any partition $P_1, \ldots, P_N$ of $\{1, \ldots, M\}$, there is a set $P_{i^\star}$ with cardinality at least $n^\star$ for some number $n^\star$ that we desire. By the generalized pigeonhole principle, it is sufficient to choose

$$M > N(n^\star - 1)$$

for this purpose. Then, to obtain a sufficient condition for the sampled observability of discrete ensembles, we show in a second step that for any $n^\star$, not necessarily consecutive, measurement times it holds that the intersection of $\ker Ce^{At_k}$ taken over these time points is trivial.

The remainder of this subsection is concerned with formulating a constructive sufficient condition for the ensemble reconstruction problem to admit a unique solution based on this idea. We first illustrate the key idea for the specific class of observable single-output systems (A, C), in which the matrix A has distinct real eigenvalues $\lambda_1, \dots, \lambda_n$.

Proposition 3.14. *Let (A, C) be an observable single-output system, where A has distinct real eigenvalues. If we are given $M > N(n-1)$ periodically sampled output snapshots $\mu_{y(t_k)}$, then the initial state μ_0 of the ensemble can be uniquely reconstructed.*

Proof. For the considered system class, we can perform a change of basis to obtain the new description $(\tilde{A}, \tilde{C})$, where $\tilde{A}$ is a diagonal matrix with the pairwise distinct eigenvalues on the diagonal and where all the entries $\tilde{c}_i$ of $\tilde{C}$ are necessarily non-zero due to observability of $(\tilde{A}, \tilde{C})$. We consider for a fixed $\Delta T > 0$ the periodic sampling scheme $t_k = (k-1)\Delta T$, where $k \in \{1, \dots, M\}$, and $M > N(n-1)$. Notice also that for the considered class of systems every periodic sampling scheme is non-pathological. With this periodic sampling scheme we obtain

$$\begin{pmatrix} \tilde{C}e^{\tilde{A}t_1} \\ \vdots \\ \tilde{C}e^{\tilde{A}t_M} \end{pmatrix} = \begin{pmatrix} 1 & \cdots & 1 \\ e^{\lambda_1 \Delta T} & \cdots & e^{\lambda_n \Delta T} \\ \vdots & & \vdots \\ e^{(M-1)\lambda_1 \Delta T} & \cdots & e^{(M-1)\lambda_n \Delta T} \end{pmatrix} \begin{pmatrix} \tilde{c}_1 & & \\ & \ddots & \\ & & \tilde{c}_n \end{pmatrix}, \tag{3.10}$$

where the first matrix on the right-hand side is a Vandermonde matrix generated by the positive and distinct real numbers $e^{\lambda_1 \Delta T}, \dots, e^{\lambda_n \Delta T}$.

Recall that since we have more than $N(n-1)$ measurement instances, for any partition $P_1, \dots, P_N$ of $\{1, \dots, M\}$, we have at least one set $P_{i^\star}$ with $|P_{i^\star}| \geq n$. Thus by virtue of Theorem 3.9, we do not have to consider every partition but only an arbitrary choice of n rows of the matrix on the left-hand side of (3.10). If any choice of n rows yields linearly independent vectors, then the reconstruction problem for a discrete ensemble will admit a unique solution. Instead of considering the matrix on the left-hand side of (3.10), we can also consider the Vandermonde matrix on the right-hand side. This is because the diagonal matrix of the entries of $\tilde{C}$ is invertible.

Now, we recall that the considered Vandermonde matrix is generated by positive and distinct real numbers $e^{\lambda_1 \Delta T}, \dots e^{\lambda_n \Delta T}$, which leads to the fact that every choice of n rows will yield linearly independent row vectors. To this end we recall the fact that for every infinite Vandermonde "matrix"

$$V(x_1, \dots, x_n) = \begin{pmatrix} 1 & \cdots & 1 \\ x_1 & \cdots & x_n \\ x_1^2 & \cdots & x_n^2 \\ \vdots & & \vdots \end{pmatrix},$$

generated by positive and distinct real numbers $x_1, \dots, x_n$, any choice of n rows will be linearly independent.

To see this, suppose that

$$\sum_{j=1}^{n} \alpha_{k_j} \begin{pmatrix} x_1^{k_j} & \dots & x_n^{k_j} \end{pmatrix} = 0,$$

so that $x_1, \dots, x_n$ are distinct positive zeros of the polynomial $p(x) = \sum_{j=1}^{n} \alpha_{k_j} x^{k_j}$. But since this polynomial $p(x)$ consists of n terms, it can have at most $n-1$ positive zeros, which is a direct consequence of *Descartes' rule of signs*. Thus, the polynomial must be trivial and the linear independence follows. □

Proposition 3.14 shows that for the specific class of observable single-output systems with distinct real eigenvalues, the combinatorial description given in Theorem 3.9 can be made effective in the sense that an explicit bound on the number of sampling times can be derived. An important fact was that for the aforementioned class of systems, any n not necessarily periodic sampling times render the sampled system observable.

In the following, we show how this idea is generalized to arbitrary observable systems (A, C) and arbitrary sampling schemes. First of all, by the pigeonhole principle, we ultimately only need to provide a number $n^\star \in \mathbb{N}$ such that an arbitrary observable system (A, C) sampled at any $n^\star$ times is observable. Instead of considering this sampled observability in terms of the matrix $(Ce^{At_{k_j}})_{j\in\{1,\dots,n^\star\}}$ having full column rank, we can also think of choosing $n^\star$ so large so that for any non-zero $h \in \mathbb{R}^n$, one of the $n^\star$ time points is not a zero of the signal $t \mapsto Ce^{At}h$. The latter viewpoint is thus related to bounds on the maximal number of zeros of exponential polynomials in a given interval, such as given in Wang et al. (2011). In the following, we give a slightly more detailed version of such a result. After establishing this result, we proceed with formulating and proving the general sampled observability result for discrete linear ensembles.

Let $\lambda_1, \dots, \lambda_q$, where $q \leq n$, denote the pairwise distinct eigenvalues of A, and d_i denote the index of λ_i as an eigenvalue of A; the dimension of the largest Jordan block associated to λ_i. Furthermore, let $\mathrm{Im}(z)$ denote the imaginary part of $z \in \mathbb{C}$.

Theorem 3.15. *Let (A, C) be observable and $h \in \mathbb{R}^n \setminus \{0\}$. Denote by d_i the index of λ_i as an eigenvalue of A. The number M_0 of zeros of the signal $t \mapsto Ce^{At}h$ in the interval $[0, T] \subset \mathbb{R}$ is bounded by*

$$M_0 \leq M_T^\star(A) := d(A) - 1 + \frac{T}{2\pi}\Delta(A),$$

where $d = \sum_{i=1}^{q} d_i$ and $\Delta = \max_{1\leq i,j\leq q} \mathrm{Im}(\lambda_i - \lambda_j)$. Furthermore, for $M \in \mathbb{N}$ with $M > M_T^\star$ the system sampled at any distinct time points $t_1, \dots, t_M$ in $[0, T]$ is observable.

In view of this result, let $M_T(A)$ denote the smallest natural number which exceeds the value $M_T^\star(A)$ throughout the remainder of this chapter.

Theorem 3.15 will be established by employing a bound for the number of zeros of so-called exponential polynomials in a given interval. First of all, we recall that any entry of e^{At}, for a matrix $A \in \mathbb{R}^{n\times n}$ with pairwise distinct eigenvalues $\lambda_1, \dots, \lambda_q$, is a linear combination of terms of the form $e^{\lambda_i t}, te^{\lambda_i t}, \dots, t^{n-1}e^{\lambda_i t}$, for some $i \in \{1, \dots, q\}$.

The (non)appearance of a term associated with a monomial in t of degree j is dictated by the index of the eigenvalue λ_i. More specifically, for an eigenvalue λ_i, only terms with monomials up to an order of the index of λ_i can appear.

In the most basic case that the considered system has a scalar output, the problem boils down to the study of the zeros of the scalar signal $y : \mathbb{R} \to \mathbb{R}$ which is the resulting output of the single-output system with an arbitrary non-zero initial state h. The scalar signal is, as reviewed in the beginning of this section, given as a linear combination of terms $e^{\lambda_i t}, te^{\lambda_i t}, \ldots, t^{n-1}e^{\lambda_i t}$, and we can write $y(t) = \sum_{i=1}^{q} P_i(t)e^{\lambda_i t}$, where $P_i \in \mathbb{R}[t]$, see also Helmke et al. (2013). Such linear combinations are known as the aforementioned exponential polynomials, or also as Bohl functions (Trentelman et al., 2001).

A general exponential polynomial is a function of the form

$$f(z) = \sum_{i=1}^{q} P_i(z)e^{\lambda_i z} \tag{3.11}$$

where $P_i \in \mathbb{C}[z]$ is a polynomial with complex coefficients in the indeterminate z, and $\lambda_i \in \mathbb{C}$. One of the main results in Voorhoeve (1976) gives an explicit bound on the number of zeros in a given interval of the real line that such an exponential polynomial f can have at most.

Lemma 3.16 (Voorhoeve (1976)). *The number M_0 of zeros of $f(z) = \sum_{i=1}^{q} P_i(z)e^{\lambda_i z}$ in the interval $[a, b] \subset \mathbb{R}$ satisfies*

$$M_0 \leq d - 1 + (b - a)\Delta/2\pi \tag{3.12}$$

where $d = \sum_{i=1}^{q}(1 + \deg P_i)$ and $\Delta = \max_{1\leq i,j\leq q} \operatorname{Im}(\lambda_i - \lambda_j)$.

We can now prove the general sampled observability result for linear systems (A, C).

Proof of Theorem 3.15. Take an arbitrary non-zero $h \in \mathbb{R}^n$. Due to observability of (A, C) at least one component of $t \mapsto Ce^{At}h$ is a non-zero signal, i.e. there exists an $i \in \{1, \ldots, m\}$ such that the signal $t \mapsto C_i e^{At}h$, where C_i denotes the ith row of the matrix C, is not identically zero. This scalar signal $t \mapsto C_i e^{At}h$ is thus a non-zero exponential polynomial of the form (3.11). By virtue of Lemma 3.16, the number of zeros M_0^i of this signal is bounded by (3.12). Clearly the zeros of $t \mapsto Ce^{At}h$ are also zeros of $t \mapsto C_i e^{At}h$, and thus $M_0 \leq M_0^i$. Lastly, since the considered exponential polynomial is generated by a linear system $\dot{x} = Ax$, the term $1 + \deg P_i$ is precisely the index of the eigenvalue λ_i of the system matrix A. □

Given this general non-pathological sampling result, we can finally state a sufficient condition for the sampled observability of discrete ensembles, which is both explicit and constructive.

Theorem 3.17. *Let (A, C) be observable. If we are given $M \geq N\, M_T(A)$ output snapshots $\mu_{y(t_k)}$ during the interval $[0, T]$, then the initial state μ_0 of the ensemble can be uniquely reconstructed.*

Proof. We show that under the given assumptions we can verify the conditions required by Theorem 3.9, i.e. that for any partition of the set $\{1, \dots, M\}$ into $P_1, \dots, P_N$, there exists one $P_{i^\star}$ such that $(Ce^{At_k})_{k \in P_{i^\star}}$ has full column rank.

Since $M \geq N\, M_T(A) > N(M_T(A) - 1)$, it follows from the pigeonhole principle that for any partition there is one $P_{i^\star}$ with $|P_{i^\star}| = M_T(A)$. By virtue of the assumption that (A, C) sampled at any $M_T(A)$ time points in the interval $[0, T]$ is observable, it immediately follows that $(Ce^{At_k})_{k \in P_{i^\star}}$ has full column rank. □

The result given by Theorem 3.17 can be succinctly phrased as the condition that for an ensemble of N systems subject to anonymization, we need to provide N times the number of sampling times that we would need to provide for the sampled observability of just one system.

One should also mention how Theorem 3.17 can in fact be concluded in a slightly more direct way by considering as the starting point the richness condition that for any family of non-zero vectors $h^{(1)}, \dots, h^{(N)} \in \mathbb{R}^n$ we have

$$\exists k \in \{1, \dots, M\}\ \forall i \in \{1, \dots, N\} \quad Ce^{At_k} h^{(i)} \neq 0.$$

From a sampled-data theoretic perspective, the problem of guaranteeing the richness condition may be understood as the task of providing as many sampling times $t_1, \dots, t_M$ so that it can be guaranteed that one of these sampling times is not a zero of any of the N non-zero signals $t \mapsto Ce^{At} h^{(i)}$ with $i \in \{1, \dots, N\}$. Since we know for a given time interval $[0, T]$ a bound on the number of zeros of a single such signal $t \mapsto Ce^{At} h^{(i)}$, using again the pigeonhole principle, we get a more direct and perhaps more intuitive answer as to why the result holds.

As we already mentioned, the considered sampled observability problem for discrete ensembles is of specific interest, also due to its direct connection to the ensemble observability problem treated in the previous chapter. In the remainder of this chapter, we proceed with studying more general ensemble observability problems that involve e.g. heterogeneous or interconnected systems. The discrete framework considered in this chapter is particularly convenient for dealing with these systems.

Sampled Observability of Discrete Heterogeneous Ensembles

In this section, we generalize our results by considering the sampled observability problem for discrete heterogeneous ensembles. By a heterogeneous ensemble, we refer to an ensemble of the form

$$\begin{aligned} \dot{x}^{(i)} &= A^{(i)} x^{(i)}, \quad x^{(i)}(0) = x_0^{(i)}, \\ y^{(i)} &= C^{(i)} x^{(i)}, \end{aligned}$$

i.e. the system and output matrices are no longer necessarily identical among the ensemble. To obtain a non-pathological sampling result here, we will extend the scope of the purely sampled-data theoretic approach that we discussed at the end of the previous subsection. The idea to approach the sampled observability problem for heterogeneous ensembles is to start from a condition similar to the richness condition in Theorem 3.1, cf. also the sampled-data theoretic interpretation given after its proof.

Suppose that measurement time points $t_1, \dots, t_M$ are chosen such that for all $x \in \mathbb{R}^n$ with $x \neq x_0^{(i)}, i \in \{1, \dots, N\}$, the condition

$$\forall j \in \{1, \dots, N\}\ \exists k \in \{1, \dots, M\}\ \forall i \in \{1, \dots, N\} \quad C^{(j)} e^{A^{(j)} t_k} x \neq C^{(i)} e^{A^{(i)} t_k} x_0^{(i)}, \tag{3.13}$$

holds, then the (unordered) initial states $x_0^{(1)}, \dots, x_0^{(N)}$ of the ensemble can be uniquely reconstructed by virtue of the given output snapshots. More precisely, the condition reflects the intuitive idea that computing for all $j \in \{1, \dots, N\}$ the hypothetical output that would result from the propagation of x with the dynamics of the jth system, we always find a snapshot at time t_k so that this hypothetical output does not match any actual output $C^{(i)} e^{A^{(i)} t_k} x_0^{(i)}$ in the snapshot. Therefore, under the condition that the available time points $t_1, \dots, t_M$ satisfy the condition given in (3.13), all points $x \in \mathbb{R}^n$ which are not equal to any of the actual initial states $x_0^{(i)}$ can indeed be identified as such from observing only the output snapshots.

In the following, we work towards formulating more explicit, i.e. verifiable, sufficient conditions for (3.13) to hold. First of all, we attempt to rewrite (3.13) by introducing the matrices

$$\tilde{A}_{ij} := \begin{pmatrix} A^{(i)} & \\ & A^{(j)} \end{pmatrix}, \quad \tilde{C}_{ij} := \begin{pmatrix} C^{(i)} & C^{(j)} \end{pmatrix}.$$

We will refer to these matrices as the system and output matrices of a so-called extended system. The situation can then be recapitulated as follows. The initial states of the ensemble can be uniquely reconstructed from the given output snapshots, if $t_1, \dots, t_M$ are chosen such that for all $x \in \mathbb{R}^n$ that satisfy $x \neq x_0^{(i)}$ for all $i \in \{1, \dots, N\}$, we have

$$\forall j \in \{1, \dots, N\}\ \exists k \in \{1, \dots, M\}\ \forall i \in \{1, \dots, N\} \quad \tilde{C}_{ij} e^{\tilde{A}_{ij} t_k} \begin{pmatrix} x_0^{(i)} \\ -x \end{pmatrix} \neq 0. \tag{3.14}$$

Although the actual $x_0^{(1)}, \dots, x_0^{(N)}$ are undisclosed to us, we can guarantee (3.14) nonetheless by the following, slightly more conservative, approach: with the definition of the "anti-diagonal"

$$\mathscr{A} := \{(x, -x) : x \in \mathbb{R}^n\} \subset \mathbb{R}^{2n},$$

we aim to provide as many $t_1, \dots, t_M$ so that for any family $\tilde{x}^{(1)}, \dots, \tilde{x}^{(N)} \in \mathbb{R}^{2n} \setminus \mathscr{A}$,

$$\forall j \in \{1, \dots, N\}\ \exists k \in \{1, \dots, M\}\ \forall i \in \{1, \dots, N\} \quad \tilde{C}_{ij} e^{\tilde{A}_{ij} t_k} \tilde{x}^{(i)} \neq 0. \tag{3.15}$$

Having arrived at this formulation, we can again view this problem as a purely sampled-data theoretic problem, analogously to what we have already discussed for the case of discrete homogeneous ensembles. In this sampled-data theoretic problem, we need to provide as many sampling times $t_1, \dots, t_M$, so that (3.15) is fulfilled, i.e. so that we always have one measurement time which avoids all the zeros of all different mappings

$$t \mapsto \tilde{C}_{ij} e^{\tilde{A}_{ij} t} \tilde{x}^{(i)}. \tag{3.16}$$

Assuming that the above exponential polynomials are non-trivial, we know that these have isolated zeros. The idea is to provide sufficiently many measurement times for the "worst case", i.e. when $j \in \{1, \dots, N\}$ is such that the system matrix $\tilde{A}_{ij}$ is able to produce the most number of zeros, and then to consider the union of zeros over all different $i \in \{1, \dots, N\}$. Then, having at hand so many measurement times so as to exceed the total number of zeros, we can fulfill (3.15).

Applying the arguments articulated above, we arrive at a non-pathological sampling result for discrete heterogeneous ensembles.

Theorem 3.18. *Consider an ensemble of observable systems $(A^{(j)}, C^{(j)})$ and suppose that for all pairs $i, j \in \{1, \dots, N\}$, the unobservable subspace of the extended system $(\tilde{A}_{ij}, \tilde{C}_{ij})$ is contained in the subspace $\mathscr{A} = \{(x, -x) : x \in \mathbb{R}^n\}$. Let $\tilde{M}_T \in \mathbb{N}$ satisfy*

$$\tilde{M}_T > 2n - 1 + \frac{T}{2\pi} \max_{i \in \{1,\dots,N\}} \Delta(A^{(i)}).$$

If we are given $M \geq N\tilde{M}_T$ output snapshots during the interval $[0, T]$, then we can uniquely reconstruct the set of initial states of the heterogeneous ensemble.

Proof. The condition that for all $i, j \in \{1, \dots, N\}$, the unobservable subspace of $(\tilde{A}_{ij}, \tilde{C}_{ij})$ is contained in $\mathscr{A}$ guarantees that the exponential polynomials (3.16) are non-trivial for all $\tilde{x}^{(i)} \in \mathbb{R}^{2n} \setminus \mathscr{A}$. This guarantees that the considered exponential polynomials have isolated zeros. The idea now is to find for a fixed $j \in \{1, \dots, N\}$, an upper bound for the total number of zeros of all the different mappings given by the fixed $j \in \{1, \dots, n\}$ and all $i \in \{1, \dots, N\}$. Taking the maximum of this bound over $j \in \{1, \dots, N\}$ will then ensure the satisfaction of (3.15).

Let $j \in \{1, \dots, N\}$ be arbitrary but fixed. Then a bound $M(j)$ for the number of zeros of one of the N signals $t \mapsto \tilde{C}_{ij} e^{\tilde{A}_{ij} t} \tilde{x}^{(i)}, i \in \{1, \dots, n\}$, with fixed $j \in \{1, \dots, N\}$, is given by

$$M(j) \geq 2n - 1 + \frac{T}{2\pi} \max_{i \in \{1,\dots,N\}} \Delta(A^{(i)}).$$

This is obtained from the definition of $M_T^\star(\tilde{A}_{ij})$ by overestimating the value $\sum_{i=1}^q d_i$ for the extended system with $2n$ and furthermore from exploiting the block diagonal structure of the matrices $\tilde{A}_{ij}$. Since $M(j)$ is in fact independent of $j \in \{1, \dots, N\}$, the result follows. □

Even though we do not exactly recover Theorem 3.17 from Theorem 3.18, the constant $2n - 1$ being slightly greater than $\sum_{i=1}^q d_i$, by inserting the fact that $A^{(j)} = A$ and $C^{(j)} = C$ in an earlier point of the ansatz for proving Theorem 3.18 we obtain the bound $M(j) \geq M_T^\star(A)$. This is because the pairwise distinct eigenvalues of the considered block diagonal matrix $\tilde{A}_{ij}$ with identical blocks A are the pairwise distinct eigenvalues of A. The resulting bound is thus the same as that obtained from Theorem 3.17, cf. Theorem 3.15. Thus we may conclude that the approach for the heterogeneous case essentially does not introduce any additional conservatism for the bound of the number of measurement times, and that, moreover, Theorem 3.18 generalizes the results from the homogeneous case in a favorable way.

Sampled Observability of Discrete Interconnected Ensembles

It is of interest to extend the results for the dynamically decoupled case considered so far to cases which allow for interconnections between the N systems in the ensemble. Furthermore, it is noted that it is particularly attractive to address such an extension in the discrete case, as there is no particular difficulty in considering such interconnections. In contrast, in the case of a continuous ensemble written as

$$\begin{aligned} \dot{x}(t) &= Ax(t), \quad x(0) \sim \mathbb{P}_0, \\ y(t) &= Cx(t) \end{aligned}$$

with a continuous distribution $\mathbb{P}_0$, as considered in the previous chapter, it is not as easy to incorporate interconnections between the systems in the ensemble. Indeed, it is difficult to speak of interconnections between individual systems as there are no individual systems that we can refer to in the first place. In the continuous problem setup, one would thus have to describe such interconnections in terms of more general partial differential equations which are beyond the scope of this thesis.

In this section, we identify one particular class of coupled ensembles of which the treatment can be reduced to the decoupled case. It will turn out that this particular class is closely related to the fundamental premise in ensemble control that is dual to the premise of anonymization, i.e. the premise of broadcast input signals. It is noted that formulating a non-pathological sampling result for the most general case of arbitrary heterogeneous systems with arbitrary interconnections appears to be rather difficult and is thus not further discussed.

The considered system class and the non-pathological sampling result are presented in the following theorem.

Theorem 3.19. *Let A, C and G be such that (A, C) and $(A - NG, C)$ are observable. Consider an ensemble of N systems described by*

$$\dot{x}^{(i)} = Ax^{(i)} + \sum_{j=1}^{N} G(x^{(j)} - x^{(i)}).$$

If we are given $M \geq \max\{M_T(A), NM_T(A - NG)\}$ output snapshots during the interval $[0, T]$, then the initial state μ_0 of the ensemble can be uniquely reconstructed.

Proof. A key observation is that we can rewrite

$$\begin{aligned} \dot{x}^{(i)} &= Ax^{(i)} + \sum_{j=1}^{N} G(x^{(j)} - x^{(i)}) \\ &= Ax^{(i)} - NGx^{(i)} + G\sum_{j=1}^{N} x^{(j)} = (A - NG)x^{(i)} + G\sum_{j=1}^{N} x^{(j)}, \end{aligned}$$

and furthermore view $\sum_{j=1}^{N} x^{(j)} =: w$ as a known exogenous input signal. Here we note that w is, in particular, the same for all individual systems $i \in \{1, \ldots, N\}$ and thus a "broadcast signal".

Now, treating w as a known signal, the sampled ensemble observability problem for

$$\begin{aligned} \dot{x}^{(i)} &= (A - NG)x^{(i)} + Gw, \\ y^{(i)} &= Cx^{(i)}, \end{aligned}$$

reduces to the decoupled homogeneous case given in Theorem 3.17, which is applicable if $(A - NG, C)$ is observable and $M \geq NM_T(A - NG)$. Next, we observe that the evolution of $w := \sum_{j=1}^{N} x^{(j)}$ is described by

$$\frac{d}{dt}w = \frac{d}{dt}\sum_{j=1}^{N} x^{(j)} = (A - NG)\sum_{j=1}^{N} x^{(j)} + NG\sum_{j=1}^{N} x^{(j)} = Aw,$$

i.e. the evolution of $\sum_{j=1}^{N} x^{(j)}$ depends only on the state of $\sum_{j=1}^{N} x^{(j)}$ itself. Moreover, by virtue of the output snapshots we can also obtain the quantity

$$\sum_{j=1}^{N} y^{(j)}(t_k) = Cw(t_k),$$

and thus, due to observability of (A, C) and the bound $M \geq M_T(A)$, we can indeed reconstruct the signal w. Combining these two parts concludes the proof. □

To conclude, the system class of interconnected ensembles for which similar results as in the decoupled case can be formulated is precisely the case of homogeneous ensembles subject to a broadcast signal, i.e.

$$\dot{x}^{(i)} = Ax^{(i)} + Bu^{(i)}$$

with $u^{(1)} = \cdots = u^{(N)}$. It is quite remarkable that the idea of broadcast signals, which is dual to our motivation from a conceptual point of view, does indeed appear rather naturally in the sampled observability problem for discrete ensembles as well. While the proof is clear from a mathematical point of view, there are some conceptually interesting points that are worthwhile to discuss further.

First of all, for the proof it was crucial that the state coupling is such that the resulting coupling signal is the same for all systems, which is the case for coupled ensembles of the form

$$\frac{d}{dt}\begin{pmatrix} x^{(1)} \\ \vdots \\ x^{(N)} \end{pmatrix} = \begin{pmatrix} \tilde{A} & & \\ & \ddots & \\ & & \tilde{A} \end{pmatrix}\begin{pmatrix} x^{(1)} \\ \vdots \\ x^{(N)} \end{pmatrix} + \begin{pmatrix} G^{(1)} & \dots & G^{(N)} \\ \vdots & & \vdots \\ G^{(1)} & \dots & G^{(N)} \end{pmatrix}\begin{pmatrix} x^{(1)} \\ \vdots \\ x^{(N)} \end{pmatrix}.$$

This allows to indeed view the coupling term as a broadcast input, the effect of which can be canceled out. If the coupling terms were different, then the operation of subtraction would not be well-defined since no labeling of the measurements $y^{(1)}(t_k), \dots, y^{(N)}(t_k)$ is available from the output snapshots.

Secondly, only because $G^{(1)} = \cdots = G^{(N)} = G$, we can write the dynamics of the broadcast input in the closed form

$$\frac{d}{dt}\sum_{j=1}^{N} x^{(j)} = (\tilde{A} + NG)\sum_{j=1}^{N} x^{(j)},$$
$$\sum_{j=1}^{N} y^{(j)} = C\sum_{j=1}^{N} x^{(j)}.$$

Lastly, the sampled output $\sum_{j=1}^{N} y^{(j)}(t_k)$ of the above system can indeed be computed from the corresponding output snapshots, as the sum is symmetric in the sense that the order of summation is irrelevant. Again, this is important because the output snapshots are missing information about associations.

Another observation worthwhile to discuss is that if we replace the state coupling by an output coupling, the foregoing result simplifies as follows.

Corollary 3.20. *Let (A, C) be observable. Consider N coupled systems described by*

$$\dot{x}^{(i)} = Ax^{(i)} + \sum_{j=1}^{N} \tilde{G}(y^{(j)} - y^{(i)}),$$

with $y^{(i)} = Cx^{(i)}$. If we are given $M \geq NM_T(A - N\tilde{G}C)$ output snapshots during the interval $[0, T]$, then the initial state μ_0 of the ensemble can be uniquely reconstructed.

Proof. Defining $G := \tilde{G}C$, we have

$$\dot{x}^{(i)} = Ax^{(i)} + \sum_{j=1}^{N} G(x^{(j)} - x^{(i)}) = (A - NG)x^{(i)} + \tilde{G}\sum_{j=1}^{N} y^{(j)}.$$

By adding the output measurements in every snapshot, we can compute the value $\tilde{G}\sum_{j=1}^{N} y^{(j)}$, and thus cancel out the effects of the input term. It remains only to show that $(A - N\tilde{G}C, C)$ is observable. The system $(A - N\tilde{G}C, C)$, however, can be viewed as the system (A, C) closed under static output feedback, which preserves observability. □

To summarize, in this subsection, we considered the more general case in which the dynamics of the individual systems within the ensemble are coupled. The results presented here include the case of all-to-all diffusive couplings, which are relevant in view of the ensemble control framework, and, more generally, also of couplings that result from input signals of the form

$$u^{(i)} = Kx^{(i)} + F\sum_{j=1}^{N} x^{(j)},$$

which are currently subject to increasing interest, see e.g. Madjidian and Mirkin (2014).

3.6 Summary and Discussion

In this chapter, we considered a more particular instance of the general ensemble observability problem that we introduced and studied in the previous chapter. More specifically, the continuous distribution was replaced by a discrete measure with finitely many support points, and the continuous measurement times by discrete measurement times. On the one hand, this problem can be seen as a natural, perhaps more tangible, particular case of the ensemble observability problem for linear systems. On the other hand, it may be independently viewed as a basic sampled-data theoretic problem in which the output measurements are received in an anonymized way.

The first part of this chapter was concerned with highlighting the connection to the previous chapter, and we thus introduced the familiar description of the state and output of a discrete ensemble using (discrete) measures. We illustrated the geometric idea of this description, which led to a discussion of the sampled observability problem as a discrete tomography problem, and which eventually resulted in a first sufficient condition for sampled observability of a discrete ensemble. This condition may be viewed as a richness condition of the family of mappings $(Ce^{At_k})_{k\in\{1,\dots,M\}}$, or, equivalently, of the family of subspaces $(\ker Ce^{At_k})_{k\in\{1,\dots,M\}}$. Since this property, arising within the context of abstract anonymization, seems to be rather basic, and has not, to the best knowledge of the author, been considered so far, we introduced the general property of richness of a family of subspaces of a given order by means of Definition 3.2. A family of subspaces is said to be rich of order N if for any set of N non-zero vectors there is a subspace within the family which contains none of these N non-zero vectors. The geometric intuition behind this property is that a rich family of subspaces is one in which the orientation of the subspaces is sufficiently "scattered in space". We then proceeded with providing some further reformulations of this richness property. For instance, in view of the richness characterizations of the previous chapter, we showed that the family $(\ker Ce^{At_k})_{k\in\{1,\dots,M\}}$ is rich of order N if and only if the finite union $\bigcup_{k\in\{1,\dots,M\}} \operatorname{im} Ce^{At_k}$ is not contained in a union of N hyperplanes. The second reformulation was in terms of intersections of certain subsets of the subspaces in the considered family, where the subsets over which the intersection is to be taken depends on partitions of the set $\{1,\dots,M\}$ into sets $P_1,\dots,P_N$. Since the intersection is, in particular, to be taken over smaller subsets, achieving a trivial intersection is in principle harder, providing some further intuition on the sufficient condition for the uniqueness of the reconstruction.

Due to the rather fruitful use of the moment-based framework in the study of continuous ensemble observability of linear systems, we introduced, as a starting point for a moment-based consideration in the discrete setup, a finite analogue of moments for discrete measures. The definition of these finite moments is, in the simplest case of discrete ensembles, based on the idea of associating to a discrete measure a polynomial which vanishes on all N support points of the discrete measure. This was then generalized to the multivariate case in a manner reminiscent of the Cramér-Wold device as it was also used in the foregoing chapter. Theorem 3.12 summarizes this general definition of moments of discrete measures and establishes the determinacy of discrete measures by these moments. Theorem 3.13 shows that the dynamics of these moments are governed by the first N tensor systems $(A_{[p]}, C^{[p]}), p \in \{1,\dots,N\}$.

The last part of this chapter dealt with sharp, explicit sufficient conditions for guaranteeing sampled observability of discrete ensembles, which are of interest in their own right. In particular, we presented non-pathological sampling results for discrete ensembles of linear systems, discrete ensembles of heterogeneous linear systems, and discrete ensembles of linear systems with a certain coupling. The non-pathological sampling result for the most basic case of a discrete ensemble of N linear systems states that a sampling frequency which is N times higher than the critical frequency for a single linear system (A, C) is sufficient for the sampled observability of the ensemble. Thus, this non-pathological sampling result also revealed that the observability of (A, C) alone is in fact sufficient for sampled observability of discrete ensembles, and that, in particular, the higher order tensor systems $(A_{[p]}, C^{[p]})$, though also explicitly appearing in this framework, eventually appear to be not as central as in the continuous framework.

4 Conclusions

Recent efforts in the control and observation of ensembles of dynamical systems, i.e. populations (or copies) of nearly identical systems, have brought forth new kinds of practical challenges, and have thereby also drawn attention to new fundamental concepts that lie at the very core of systems and control theory. A particularly suitable example for the latter assertion are state and parameter estimation problems for ensembles, in particular heterogeneous cell populations, which motivated the introduction and the study of ensemble observability of dynamical systems in this thesis. More specifically, in the context of state and parameter estimation for heterogeneous cell populations, one aims at inferring the distribution of initial states or parameters of a population of heterogeneous cells in order to understand the heterogeneity within the population, which, e.g. in cancer cells, is one of the key factors for the resistance of cancer to treatments. The measurement data for performing such state or parameter estimation is, however, typically obtained via high-throughput devices such as flow cytometers, which provide a vast number of output measurements of individual cells at each measurement time. Flow cytometers do not, however, allow one to specify which cell to measure beforehand, and, in fact, in many cases result in killing the cell upon taking a measurement. Therefore, one cannot make precise statements on the measured data beyond the fact that, at each measurement time, a vast number of output measurements of cells are provided. An adequate mathematical model for capturing this circumstance is obtained through viewing the output measurement data provided by the flow cytometers as samples drawn from a population in a statistical sense. Upon a further idealization, the control theoretic problem at hand is the question of whether we can infer a distribution of initial states from observing the evolution of the distribution of outputs.

The first part of this thesis was concerned with studying the general problem of ensemble observability for dynamical systems, i.e. properties of dynamical systems with output which allow one to infer a continuous distribution of initial states from observing the evolution of the resulting output distribution continuously. In the second part of this thesis, we considered a slightly more specific instance of the general ensemble observability problem, in which the observability of an ensemble of finitely many linear systems from its output snapshots at discrete times is investigated. Thus, by introducing and establishing a theory of observability for ensembles as a counterpart to controllability of ensembles within this thesis, we contributed to the very basics of a control theory of ensembles. Establishing a basic theoretical foundation (centered around the key concepts of controllability and observability) is considered a necessary first step for tackling even more advanced applications related to ensembles in the future.

In the following, we summarize the main results of the thesis, discuss these in a more general context, and indicate possible directions for future research.

4.1 Summary and Discussion

In Chapter 2, we introduced and studied the ensemble observability problem for dynamical systems. We defined ensemble observability of a finite-dimensional dynamical system with output as the property that a distribution of initial states can be reconstructed from observing only the evolution of the corresponding distribution of outputs. We first studied the problem in the simplest setting of linear systems. Therein, we showed that the measured output densities of an ensemble of linear systems (A, C) at some time $t \geq 0$ can be viewed as being obtained from integrating the initial density of interest over all affine subspaces in the quotient space $\mathbb{R}^n / \ker Ce^{At}$. This connection is based on a specific geometric perspective on the ensemble observability problem that we took in Sections 2.2 and 2.3, which eventually brought together the two ideas of observability and mathematical tomography; interestingly, these two ideas emerged at approximately the same time in the 1960s, though as completely independent problems.

The connection to tomography is considered a key theme of the analysis in this thesis, which led to a first theoretical characterization for ensemble observability of linear systems with respect to a specific class of continuous probability distributions in Theorem 2.7. This sufficient condition required the evolution of the normal spaces of the affine subspaces one is integrating the unknown density over to be contained in no proper projective variety. We discussed how this may be viewed as a persistence of excitation condition for the linear system (A, C).

Motivated by the argument employing real analyticity of the characteristic function in the proof of Theorem 2.7 in the tomography-based framework, for the further analysis in Section 2.4 we chose to restrict our attention to moment-determinate distributions. Ensemble observability of a linear system with respect to this class of distributions is equivalent to the reconstructability of the moments of the initial distribution. It is imperative to note that it was the tomography-based framework, and, in particular, the consideration via the Cramér-Wold theorem which demonstrated the *necessity* to impose further regularity assumptions on the considered densities, and is furthermore the justification for solely considering moment-determinate distributions. The description of the dynamics of the moments was shown to be conveniently described in the framework of tensor systems, which led to a more general characterization for ensemble observability in Theorem 2.16. Moreover, this result yielded a fundamental duality between the two different characterizations of ensemble observability of linear systems.

As the formulation given in Theorem 2.16 requires considering the observability of infinitely many linear tensor systems, we reformulated this result for a specific class of systems, i.e. single-output systems for which the system matrix has distinct eigenvalues, which yielded a more tractable result in Theorem 2.20. For this specific class of systems, a sufficient condition for ensemble observability is a linear independence condition on the spectrum of A, which is known as a non-resonance condition in the dynamical systems literature. We also discussed the slightly more specialized case in which an independence assumption is placed on the considered initial state distribution. Therein, the geometric characterization of ensemble observability requires that $\bigcup_{t\geq 0} \operatorname{im}(Ce^{At})^\top$ is not contained in a proper algebraic variety defined by

$$a_1 x_1^p + \cdots + a_n x_n^p = 0.$$

In Section 2.5, we examined the ensemble observability problem for nonlinear systems. We illustrated that in this case a tomography-based viewpoint is still valid, though the attempt to reformulate the ensemble observability problem in terms of a standard (nonlinear) tomography problem (involving transforms that integrate along manifolds) turned out to become more involved. Nevertheless, the same geometric ideas of the first part of Chapter 2 were applicable and led to valuable insights. We discussed the generalization of the moment-based framework to the nonlinear setting on a two-dimensional nonlinear oscillator and highlighted the particular difficulties faced in the most general nonlinear case.

Chapter 3 was devoted to the study of a more particular instance of the general ensemble observability problem for linear systems, which may be summarized as the consideration of the general problem of Chapter 2 in a discrete framework. We discussed how this discrete version of the problem may be viewed as a basic sampled-data control problem, in which one considers a group of N linear systems, of which outputs are measured, but eventually only obtained in an anonymized manner. We described this sampled observability problem for discrete ensembles in terms of a discrete tomography problem, which led to a first sufficient condition in Theorem 3.1.

This characterization has the geometric interpretation that the family of subspaces $\ker Ce^{At_k}$ needs to admit sufficient movement in all directions of $\mathbb{R}^n$. More precisely, the geometric richness condition, which we defined in a more general context in Definition 3.2, may be rigorously described as separating a set of N unit vectors in $\mathbb{R}^n$. With Theorem 3.5 we also provided a reformulation in terms of the more familiar description from Chapter 2: the union $\bigcup_{k\in\{1,\ldots,M\}} \operatorname{im}(Ce^{At_k})^\top$ must not be contained in an algebraic variety defined by

$$\langle x, h^{(1)}\rangle \cdots \langle x, h^{(N)}\rangle = 0$$

with a family of non-zero vectors $h^{(1)}, \ldots, h^{(N)} \in \mathbb{R}^n$.

Section 3.4 was concerned with introducing a moment-based framework for the discrete setup. By employing one of the basic ideas of the thesis, namely the reduction of a multivariate problem to a simpler univariate problem, we described a way to obtain finitely many symmetric polynomial equations that characterize a discrete measure uniquely. In particular, the result summarized in Theorem 3.12 generalizes the previously considered idea of symmetric measurement equations in multitarget tracking. We then investigated the dynamics of these moments of discrete measures and recovered the same link to tensor systems as we found in Chapter 2. The general result on the moment dynamics of discrete measures is given in Theorem 3.13.

In the last part of Chapter 3, we provided a series of sharp non-pathological sampling results. The simplest result for the sampled observability of a discrete ensemble of homogeneous linear systems shows that the observability of (A, C) alone is sufficient, and moreover that it is required to sample N times faster than the critical frequency for a single linear system (A, C). This also revealed that, even though tensor systems naturally appear in the discrete setup as well, the observability of higher order tensor systems is actually irrelevant. Lastly, we extended the basic result for discrete ensembles to more general settings, such as to discrete ensembles with heterogeneous dynamics and discrete ensembles with a specific coupling between the individual systems.

To summarize, the two recurrent themes in this thesis were the natural connection to tomography problems and the use of polynomial descriptions (and, in particular, tensor systems). The connection to tomography is based on the elementary geometric interpretation of the considered (ensemble) observability problem and eventually allowed us to formulate several persistence of excitation results. Due to the modeling of the initial states of an ensemble through probability measures, discrete or continuous, it does not come as a surprise that polynomials eventually played a fundamental role in our study. This is because, e.g. in the case of continuous distributions, the moments of the distribution are expected values of monomials.

In the next section, we discuss some possible future research directions that arise from the study and results in this thesis.

4.2 Outlook

While we were able to present a rather complete picture of the ensemble observability problem for linear systems, for the case of nonlinear systems we were only able to present the general geometric idea and to prove ensemble observability in special cases. It is, however, of great interest to extend our knowledge of the nonlinear ensemble observability problem. A natural idea is to formulate the nonlinear ensemble observability problem in the framework of the Liouville equation which was already mentioned in the introduction. For certain output mappings $h : \mathbb{R}^n \to \mathbb{R}^m$, with constant m-dimensional Jacobian, the state and output densities can be described by the system

$$\begin{aligned} \frac{\partial}{\partial t} p(t,x) &= -\operatorname{div}(p(t,x)f(x)), \qquad p(0,x) = p_0(x), \\ p_{y(t)}(y) &= \int_{h^{-1}(\{y\})} p(t,x)\, dS. \end{aligned}$$

Therein, the first equation describes the evolution of the state density in a vector field $f : \mathbb{R}^n \to \mathbb{R}^n$, and the second equation specifies the output density as a marginalization of the state density over the output mapping. Given this formulation, one could attempt to apply the framework of infinite-dimensional linear systems theory, though it is noted that therein the considered output mapping is nonstandard; indeed the particular integration is a hallmark of tomography problems. The formulation of the ensemble observability problem for nonlinear systems in the framework of partial differential equations also brings us to the idea of posing the ensemble observability problem in a greater context of arbitrary population models described by more general partial differential equations. Indeed, while individual systems are dynamic in this thesis, the ensembles may be referred to as rather static in the sense that no individual systems are added to or removed from the ensemble; many ensembles in practical applications are, however, very dynamic. For example, cells in a cell population are subject to cell division / cell death and may, furthermore, exhibit couplings with other cells, either directly or through a common medium. These additional considerations all lead to a behavior which may be regarded as a more dynamic aspect of the ensemble at the population level.

Summarizing, it would also be of interest to extend the considered notion of ensemble observability to more general population models, and, in particular, to dynamic populations. The measure theoretic approach from which we set off in this thesis also appears promising for this generalization.

Furthermore, it is of great interest to extend the framework of practical reconstruction methods for the (initial) state distribution of an ensemble. A possible approach would be to establish an observer-based reconstruction technique that employs the formulation in terms of partial differential equations.

A further important extension in the study of ensemble observability is the consideration of external input signals. In this thesis, we did not discuss the explicit inclusion of inputs. While this is partly due to the fact that incorporating *broadcast* input signals in the linear ensemble observability problem is trivial (for the effect of the known input can be canceled out), input signals start to play a crucial role once nonlinear systems are considered. A particularly important example is given by ensembles of bilinear systems

$$\dot{x}(t) = Ax(t) + u(t)Bx(t),$$

which have been studied e.g. in Li and Khaneja (2006) in the context of problems in quantum control. For this bilinear setup, a broadcast input signal that is applied on the population level may still lead to a specific feedback mechanism for the individual systems due to the state-dependency introduced by the bilinear term, thus giving rise to a more complex behavior. A similar situation is expected when a combination of heterogeneities in the dynamics (not just in initial states) and pure broadcast input signals is considered.

Regarding the second part of the thesis, we did not discuss the practical state estimation problem for discrete ensembles subject to noise in the output measurements. This state estimation problem has a particularly interesting geometric interpretation in the framework of discrete tomography, in which the backprojection lines are not bound to run through the actual points perfectly, but are subject to some displacements about the actual points. This may therefore be described as a situation in which the backprojection lines are "fuzzed out", and the practical state reconstruction problem is thus equivalent to placing N points for the estimates such that a certain least squares problem is solved. Obtaining efficient methods for the practical reconstruction problem is also relevant for multitarget tracking problems.

Bibliography

N. Akhiezer. The classical moment problem and some related questions in analysis. *Hafner, New York*, 1965.

V. I. Arnol'd. *Geometrical methods in the theory of ordinary differential equations.* Springer Verlag, Berlin, New York, 1983.

J. Baillieul. Controllability and observability of polynomial dynamical systems. *Nonlinear Analysis: Theory, Methods & Applications*, 5(5):543–552, 1981.

H. T. Banks, Z. R. Kenz, and W. C. Thompson. A review of selected techniques in inverse problem nonparametric probability distribution estimation. *Journal of Inverse and Ill-Posed Problems*, 20(4):429–460, 2012.

Y. Bar-Shalom. Tracking methods in a multitarget environment. *IEEE Transactions on Automatic Control*, 23(4):618–626, 1978.

A. Becker and T. Bretl. Approximate steering of a unicycle under bounded model perturbation using ensemble control. *IEEE Transactions on Robotics*, 28(3):580–591, 2012.

A. Becker, C. Onyuksel, T. Bretl, and J. McLurkin. Controlling many differential-drive robots with uniform control inputs. *International Journal of Robotics Research*, 33 (13):1626–1644, 2014.

H. Blom and E. Bloem. Probabilistic data association avoiding track coalescence. *IEEE Transactions on Automatic Control*, 45(2):247–259, 2000.

L. Boltzmann. Einige allgemeine Sätze über Wärmegleichgewicht. *Wiener Berichte*, 63: 679–711, 1871.

R. W. Brockett. Lie algebras and Lie groups in control theory. In D. Q. Mayne and R. W. Brockett, editors, *Geometric Methods in System Theory*, volume 3 of *NATO Advanced Study Institutes Series*, pages 43–82. Springer, Netherlands, 1973.

R. W. Brockett. Optimal control of the Liouville equation. *AMS IP Studies in Advanced Mathematics*, 39:23 – 35, 2007.

R. W. Brockett. On the control of a flock by a leader. *Proc. of the Steklov Institute of Mathematics*, 268(1):49–57, 2010.

R. W. Brockett. Notes on the control of the Liouville equation. In P. Cannarsa and J. M. Coron, editors, *Control of Partial Differential Equations*, pages 101–129. Springer, Berlin-Heidelberg, 2012.

R. W. Brockett. The early days of geometric nonlinear control. *Automatica*, 50(9): 2203–2224, 2014.

R. W. Brockett and N. Khaneja. On the stochastic control of quantum ensembles. In *System Theory*, pages 75–96. Springer, 2000.

T. Carleman. Application de la théorie des équations intégrales linéaires aux systémes d'équations différentielles non linéaires. *Acta Mathematica*, 59(1):63–87, 1932.

K.-C. Chang and Y. Bar-Shalom. Joint probabilistic data association for multitarget tracking with possibly unresolved measurements and maneuvers. *IEEE Transactions on Automatic Control*, 29(7):585–594, 1984.

A. M. Cormack. Representation of a function by its line integrals, with some radiological applications. *Journal of Applied Physics*, 34(9):2722–2727, 1963.

A. M. Cormack. Representation of a function by its line integrals, with some radiological applications. II. *Journal of Applied Physics*, 35(10):2908–2913, 1964.

H. Cramér and H. Wold. Some theorems on distribution functions. *Journal of the London Mathematical Society*, 1(4):290–294, 1936.

W. Dayawansa and C. Martin. Observing linear dynamics with polynomial output functions. *Systems & Control Letters*, 9(2):141–148, 1987.

A. Einstein. Kinetische Theorie des Wärmegleichgewichtes und des zweiten Hauptsatzes der Thermodynamik. *Annalen der Physik*, 314(10):417–433, 1902.

W. Feller. An introduction to probability and its applications, vol. II. *Wiley, New York*, 1971.

P. Fuhrmann. On controllability and observability of systems connected in parallel. *IEEE Transactions on Circuits and Systems*, 22(1):57–57, 1975.

J. W. Gibbs. *Elementary principles in statistical mechanics*. Yale University Press, New Haven, 1902.

R. Gordon, R. Bender, and G. T. Herman. Algebraic Reconstruction Techniques (ART) for three-dimensional electron microscopy and X-ray photography. *Journal of Theoretical Biology*, 29(3):471–481, 1970.

J. Hasenauer, S. Waldherr, M. Doszczak, N. Radde, P. Scheurich, and F. Allgöwer. Identification of models of heterogeneous cell populations from population snapshot data. *BMC Bioinformatics*, 12(1):125, 2011a.

J. Hasenauer, S. Waldherr, M. Doszczak, P. Scheurich, N. Radde, and F. Allgöwer. Analysis of heterogeneous cell populations: a density-based modeling and identification framework. *Journal of Process Control*, 21(10):1417–1425, 2011b.

S. Helgason. *Integral Geometry and Radon Transforms*. Springer, New York, 2011.

U. Helmke and M. Schönlein. Uniform ensemble controllability for one-parameter families of time-invariant linear systems. *Systems & Control Letters*, 71:69–77, 2014.

U. Helmke, K. Huper, and M. Khammash. Global identifiability of a simple linear model for gene expression analysis. In *Proc. 52nd IEEE Conference on Decision and Control*, pages 7149–7154, 2013.

L. A. Herzenberg, J. Tung, W. A. Moore, L. A. Herzenberg, and D. R. Parks. Interpreting flow cytometry data: a guide for the perplexed. *Nature immunology*, 7(7): 681–685, 2006.

S. Kaczmarz. Angenäherte Auflösung von Systemen linearer Gleichungen. *Bulletin International de l'Academie Polonaise des Sciences et des Lettres*, 35:355–357, 1937.

R. E. Kalman. On the general theory of control systems. In *Proc. First International Congress of Automatic Control*, pages 481–491, 1960.

R. E. Kalman. Mathematical description of linear dynamical systems. *SIAM Journal on Applied Mathematics*, 1(2):152–192, 1963.

E. Kamen. Multiple target tracking based on symmetric measurement equations. *IEEE Transactions on Automatic Control*, 37(3):371–374, 1992.

E. Kamen and C. Sastry. Multiple target tracking using products of position measurements. *IEEE Transactions on Aerospace and Electronic Systems*, 29(2):476–493, 1993.

F. Keinert. Inversion of k-plane transforms and applications in computer tomography. *SIAM Review*, 31(2):273–298, 1989.

P. Kingston and M. Egerstedt. Index-free multiagent systems: An eulerian approach. In *Proc. 2nd IFAC Workshop on Estimation and Control of Networked Systems (NecSys)*, pages 215–220, 2010.

A. Lasota and M. Mackey. *Chaos, Fractals, and Noise: Stochastic Aspects of Dynamics.* Springer, 1994.

J.-S. Li. Ensemble control of finite-dimensional time-varying linear systems. *IEEE Transactions on Automatic Control*, 56(2):345–357, 2011.

J.-S. Li and N. Khaneja. Control of inhomogeneous quantum ensembles. *Physical Review A*, 73(3):030302, 2006.

J.-S. Li and N. Khaneja. Ensemble control of Bloch equations. *IEEE Transactions on Automatic Control*, 54(3):528–536, 2009.

D. G. Luenberger. *Optimization by Vector Space Methods.* Wiley, New York, 1969.

D. Madjidian and L. Mirkin. Distributed control with low-rank coordination. *IEEE Transactions on Control of Network Systems*, 1(1):53–63, 2014.

A. Markoe. *Analytic Tomography, Encyclopedia of Mathematics and its Applications.* Cambridge University Press, 2006.

J. C. Maxwell. On Boltzmann's theorem on the average distribution of energy in a system of material points. *Transactions of the Cambridge Philosophical Society*, 12: 547–570, 1879.

A. Milias-Argeitis, S. Summers, J. Stewart-Ornstein, I. Zuleta, D. Pincus, H. El-Samad, M. Khammash, and J. Lygeros. In silico feedback for in vivo regulation of a gene expression circuit. *Nature biotechnology*, 29(12):1114–1116, 2011.

F. Natterer. *The Mathematics of Computerized Tomography.* Wiley, New York, 1986.

J. Radon. Über die Bestimmung von Funktionen durch ihre Integralwerte längs gewisser Mannigfaltigkeiten. *Berichte Sächsischer Akademie der Wissenschaften*, 69:262–277, 1917.

R. Rajaram, U. Vaidya, M. Fardad, and B. Ganapathysubramanian. Stability in the almost everywhere sense: A linear transfer operator approach. *Journal of Mathematical Analysis and Applications*, 368(1):144–156, 2010.

A. Rantzer. A dual to Lyapunov's stability theorem. *Systems & Control Letters*, 42(3): 161–168, 2001.

A. Rényi. On projections of probability distributions. *Acta Mathematica Hungarica*, 3 (3):131–142, 1952.

S. Sastry. *Nonlinear systems: analysis, stability, and control*, volume 10. Springer, New York, 1999.

I. R. Shafarevich and K. A. Hirsch. *Basic algebraic geometry*, volume 1. Springer, New York, 1977.

H. Sira-Ramirez. Algebraic condition for observability of non-linear analytic systems. *International Journal of Systems Science*, 19(11):2147–2155, 1988.

P. Smith and G. Buechler. A branching algorithm for discriminating and tracking multiple objects. *IEEE Transactions on Automatic Control*, 20(1):101–104, 1975.

H. Trentelman, A. A. Stoorvogel, and M. Hautus. *Control theory for linear systems.* Springer, London, 2001.

M. Voorhoeve. On the oscillation of exponential polynomials. *Mathematische Zeitschrift*, 151(3):277–294, 1976.

S. Waldherr, S. Zeng, and F. Allgöwer. Identifiability of population models via a measure theoretical approach. In *Proc. 19th IFAC World Congress*, pages 1717 – 1722, 2014.

L. Y. Wang, C. Li, G. Yin, L. Guo, and C.-Z. Xu. State observability and observers of linear-time-invariant systems under irregular sampling and sensor limitations. *IEEE Transactions on Automatic Control*, 56(11):2639–2654, 2011.

Y. Wang and F. J. Doyle. Reachability of particle size distribution in semibatch emulsion polymerization. *AIChE Journal*, 50(12):3049–3059, 2004.

N. Wiener. The homogeneous chaos. *American Journal of Mathematics*, 60(4):897–936, 1938.

C. Zechner, J. Ruess, P. Krenn, S. Pelet, M. Peter, J. Lygeros, and H. Koeppl. Moment-based inference predicts bimodality in transient gene expression. *Proceedings of the National Academy of Sciences*, 109(21):8340–8345, 2012.

S. Zeng and F. Allgöwer. On the ensemble observability problem for nonlinear systems. In *Proc. 54th IEEE Conference on Decision and Control*, pages 6318 – 6323, 2015.

S. Zeng, S. Waldherr, and F. Allgöwer. An inverse problem of tomographic type in population dynamics. In *Proc. 53rd IEEE Conference on Decision and Control*, pages 1643–1648, 2014.

S. Zeng, H. Ishii, and F. Allgöwer. On the state estimation problem for discrete ensembles from discrete-time output snapshots. In *Proc. American Control Conference*, pages 4844 – 4849, 2015a.

S. Zeng, H. Ishii, and F. Allgöwer. Sampled observability of discrete heterogeneous ensembles from anonymized output measurements. In *Proc. 54th IEEE Conference on Decision and Control*, pages 5683 – 5688, 2015b.

S. Zeng, H. Ishii, and F. Allgöwer. Sampled observability and state estimation of discrete linear ensembles. *IEEE Transactions on Automatic Control*, 2016a. Conditionally accepted.

S. Zeng, S. Waldherr, C. Ebenbauer, and F. Allgöwer. Ensemble observability of linear systems. *IEEE Transactions on Automatic Control*, 61(6):1452–1465, 2016b.